Titles in This Series

81 **Steven G. Krantz,** Geometric analysis and function

80 **Vaughan F. R. Jones,** Subfactors and knots, 1991

79 **Michael Frazier, Björn Jawerth, and Guido Weiss,** Littlewood-Paley theory and the study of function spaces, 1991

78 **Edward Formanek,** The polynomial identities and variants of $n \times n$ matrices, 1991

77 **Michael Christ,** Lectures on singular integral operators, 1990

76 **Klaus Schmidt,** Algebraic ideas in ergodic theory, 1990

75 **F. Thomas Farrell and L. Edwin Jones,** Classical aspherical manifolds, 1990

74 **Lawrence C. Evans,** Weak convergence methods for nonlinear partial differential equations, 1990

73 **Walter A. Strauss,** Nonlinear wave equations, 1989

72 **Peter Orlik,** Introduction to arrangements, 1989

71 **Harry Dym,** J contractive matrix functions, reproducing kernel Hilbert spaces and interpolation, 1989

70 **Richard F. Gundy,** Some topics in probability and analysis, 1989

69 **Frank D. Grosshans, Gian-Carlo Rota, and Joel A. Stein,** Invariant theory and superalgebras, 1987

68 **J. William Helton, Joseph A. Ball, Charles R. Johnson, and John N. Palmer,** Operator theory, analytic functions, matrices, and electrical engineering, 1987

67 **Harald Upmeier,** Jordan algebras in analysis, operator theory, and quantum mechanics, 1987

66 **G. Andrews,** q-Series: Their development and application in analysis, number theory, combinatorics, physics and computer algebra, 1986

65 **Paul H. Rabinowitz,** Minimax methods in critical point theory with applications to differential equations, 1986

64 **Donald S. Passman,** Group rings, crossed products and Galois theory, 1986

63 **Walter Rudin,** New constructions of functions holomorphic in the unit ball of C^n, 1986

62 **Béla Bollobás,** Extremal graph theory with emphasis on probabilistic methods, 1986

61 **Mogens Flensted-Jensen,** Analysis on non-Riemannian symmetric spaces, 1986

60 **Gilles Pisier,** Factorization of linear operators and geometry of Banach spaces, 1986

59 **Roger Howe and Allen Moy,** Harish-Chandra homomorphisms for $\mathfrak{p}$-adic groups, 1985

58 **H. Blaine Lawson, Jr.,** The theory of gauge fields in four dimensions, 1985

57 **Jerry L. Kazdan,** Prescribing the curvature of a Riemannian manifold, 1985

56 **Hari Bercovici, Ciprian Foiaş, and Carl Pearcy,** Dual algebras with applications to invariant subspaces and dilation theory, 1985

55 **William Arveson,** Ten lectures on operator algebras, 1984

54 **William Fulton,** Introduction to intersection theory in algebraic geometry, 1984

53 **Wilhelm Klingenberg,** Closed geodesics on Riemannian manifolds, 1983

52 **Tsit-Yuen Lam,** Orderings, valuations and quadratic forms, 1983

51 **Masamichi Takesaki,** Structure of factors and automorphism groups, 1983

50 **James Eells and Luc Lemaire,** Selected topics in harmonic maps, 1983

49 **John M. Franks,** Homology and dynamical systems, 1982

48 **W. Stephen Wilson,** Brown-Peterson homology: an introduction and sampler, 1982

47 **Jack K. Hale,** Topics in dynamic bifurcation theory, 1981

46 **Edward G. Effros,** Dimensions and C^*-algebras, 1981

45 **Ronald L. Graham,** Rudiments of Ramsey theory, 1981

Geometric Analysis and Function Spaces

Conference Board of the Mathematical Sciences

CBMS

Regional Conference Series in Mathematics

Number 81

Geometric Analysis and Function Spaces

Steven G. Krantz

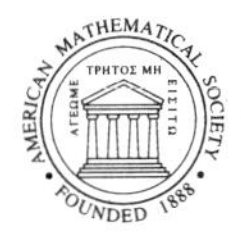

Published for the
Conference Board of the Mathematical Sciences
by the
American Mathematical Society
Providence, Rhode Island
with support from the
National Science Foundation

Expository Lectures
from the NSF-CBMS Regional Conference
held at George Mason University, Fairfax, Virginia
May 26–31, 1992

Research partially supported by
National Science Foundation Grant DMS-9108504.

1991 *Mathematics Subject Classification.* Primary 32Axx, 32Bxx, 32Exx, 32Fxx, 32Gxx, 32Hxx, 31Axx, 31Bxx, 35Nxx, 42Bxx.

Library of Congress Cataloging-in-Publication Data

Krantz, Steven G. (Steven George), 1951–
Geometric analysis and function spaces/Steven G. Krantz.
p. cm. —(Regional conference series in mathematics; no. 81)
Includes bibliographical references and index.
ISBN 0-8218-0734-X (alk. paper)
1. Geometry, Differential. 2. Functions of complex variables. I. Conference Board of the Mathematical Sciences. II. Title. III. Series.
QA1.R33 no. 81
[QA641]
510 s—dc20 93-4170
[516.3′62] CIP

Printed in the United States of America
The paper used in this book is acid-free and falls within the guidelines
established to ensure permanence and durability. ♾

This publication was typeset using $\mathcal{A}\mathcal{M}\mathcal{S}$-TEX,
the American Mathematical Society's TEX macro system.

10 9 8 7 6 5 4 3 2 1 97 96 95 94 93

To see a world in a grain of sand
And a heaven in a wild flower,
Hold infinity in the palm of your hand
And eternity in an hour.

—Wm. Blake

Table of Contents

Preface

This short book is not a mathematics monograph in the usual sense of the word. It is more of a manifesto. Its purpose is to illustrate, by way of a number of examples, an interaction between geometry and analysis. In some of these examples the geometry is of the classical Euclidean variety; in others it is Riemannian or Kähler or Finsler geometry. By the same token, some of the analysis that occurs is at the textbook level while other instances use many of the powerhouse tools from the last fifteen or twenty years. The interest is in the symbiosis.

Some of the material in this book is standard. Other parts are well known but have not been recorded. Still others are rather speculative in nature. We shall try to be specific about the places where the picture is incomplete.

In virtually every example that I treat, complex function theory (of either one or several variables) acts as a catalyst. In some instances, the function theory helps to set up or motivate an attractive problem and then quickly fades into the background. In other instances the complex analysis remains nearly to the end, being stripped away in layers to reveal geometry, analysis, and operator theory hidden in its folds.

Another unifying theme of this work will be partial differential equations. In many of the examples, either the Cauchy Riemann equations (of one or several variables), the classical Laplacian, or the Laplace-Beltrami operator for some Riemannian metric will play a role. Partial differential equations will not only serve us as a technical tool, but also as a driving force behind certain questions.

And one question that I want to ask repeatedly is this: "Where do the function spaces that we use come from? Are they still well-suited for the problems that we study today?" Function spaces are a relatively modern invention. The L^p, Hölder/Lipschitz, and Sobolev spaces have all been developed primarily in the twentieth century. These spaces behave canonically under the groups that act on Euclidean space: the translations, rotations, and dilations. And so they should, because they were developed in significant part to study the regularity properties of the classical Laplacian.

From the vantage point of the 1990's, it is safe to say that the theory of translation invariant operators on Euclidean space, and their mapping properties on the classical function spaces mentioned above, is well understood. Both in the setting of (harmonic) analysis on Euclidean space and of the function theory of several complex variables we have now turned our attention to the analysis of domains. It has become clear in the last fifteen years that the function theory of several complex variables consists in understanding how the Levi geometry of the boundary of a domain influences the holomorphic function theory on the interior. I want to suggest in this context, among others, that it is useful to set aside the classical function spaces and think in terms of new spaces *that are adapted to the geometry of the given situation.*

We shall not throw out the proverbial baby with the bath water: the new function spaces are closely linked to, and motivated by, the classical ones. In many familiar settings the new function spaces are in fact classical ones in disguise. Because we give them new descriptions, and because they are developed canonically for the given context, they will be easier and more natural to treat.

Already in the preface we have introduced a number of themes: analysis, geometry, partial differential equations, and function spaces. These will be present, woven into what we hope is a comfortable tapestry, throughout this book. Of course there is a simple explanation for this setup. It is natural for me to speak here of topics in which I have had a hand in the last twenty years. For synergistic reasons, if for none other, there are bound to be at least tenuous links among these topics. This is thus an opportunity for me to make the links more explicit and, I hope, to clarify some connections and to point in new directions.

It is a pleasure to thank F. Beatrous, F. Di Biase, S. Fu, X. Huang, I. Graham, K. T. Kim, J. McNeal, T. Nguyen, H. Parks, P. Pflug, and J. Yu for reading various drafts of the manuscript and contributing useful suggestions.

Flavia Colonna, David Singman, and Daniele Struppa did a splendid job of organizing the CBMS conference during which the lectures on which this book is based were presented. I thank them for giving me this opportunity.

—Steven G. Krantz

Chapter 1. Complex Geometry and Applications

1.1. Nonisotropic Spaces. Nonisotropic spaces arise naturally in many different contexts—often in connection with a partial differential equation. In the classical setting, the heat equation gives rise to such spaces:

Consider the (parabolic) heat equation

$$u_t = u_{xx}. \tag{1}$$

on the half-plane $\mathbb{R}^2_+$. The fundamental solution for this equation is

$$H(x,t) = \frac{1}{2\sqrt{\pi t}} \cdot e^{-x^2/(4t)}.$$

In fact H is annihilated by the heat operator and, as $t \to 0^+$, the function $H(\cdot, t)$ tends in the weak star topology of measures to the Dirac mass at the origin in $\mathbb{R}$. See [**WID**]. It follows that if f is a testing function on $\mathbb{R} = \partial\mathbb{R}^2_+$ then

$$u(x,t) = H * f = \int_{\mathbb{R}} H(x-w,t) f(w)\, dw$$

provides a solution to (1) which assumes, in a natural sense, the boundary limit f.

We can get a glimpse of the nonisotropic regularity of the heat equation by noticing that (1) mandates that one derivative in t counts the same as two derivatives in x. In particular, if $u = H * f$ is a solution of the heat equation then u_{xx} will have just the same growth and smoothness as u_t. One may formulate these statements more precisely using special function spaces. We shall not indulge ourselves in the details here, but refer the reader to the paper [**FRI**].

An elementary context in which nonisotropic analysis arises in complex function theory is as follows. Consider the unit ball B in $\mathbb{C}^2$. Let $P = (P_1, 0)$ be a point of distance δ from the boundary point $\mathbb{1} = (1,0)$. [In what follows it will be convenient to let $\delta = \delta_\Omega(P)$ denote the distance of the point P from the boundary of Ω.] Assume that $\delta < 1$. Refer to Figure 1.1. Then trivial algebra shows that the points of the discs

$$\mathbf{d}_n = \{(w_1, 0) : |w_1 - P_1| < \delta\}$$

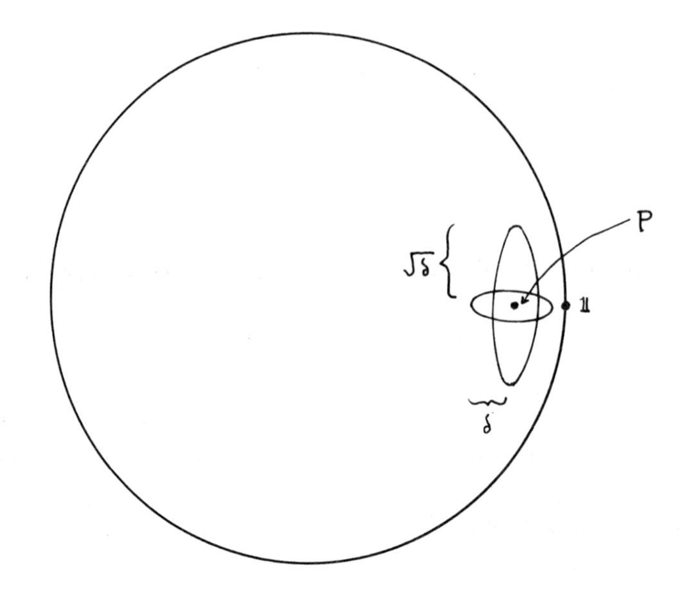

FIGURE 1.1

and

$$\mathbf{d}_t = \{(P_1, w_2) : |w_2| < \sqrt{\delta}\}$$

lie in B.

Let f be a bounded holomorphic function on B : say that $|f| \leq M$. Consider the restrictions of f to $\mathbf{d}_n$ and to $\mathbf{d}_t$. Applying the classical one variable Cauchy estimates to each of these restrictions, we find that

$$\left|\frac{\partial f}{\partial z_1}(P)\right| \leq \frac{M}{\delta} \tag{C_1}$$

and

$$\left|\frac{\partial f}{\partial z_2}(P)\right| \leq \frac{M}{\sqrt{\delta}}. \tag{C_2}$$

Thus we see that, in a palpable sense, tangential derivatives seem to count only half as much as normal derivatives—at least for holomorphic functions.

Of course there is nothing special about the domain B in the last calculation. For if Ω is any smoothly bounded domain in $\mathbb{C}^n$ (C^2 boundary will suffice) and if f is a holomorphic function on Ω satisfying $|f| \leq M$ then we may perform a similar analysis just by noting the following: If $Q \in \partial\Omega$ then there is an internally tangent ball $B_Q \equiv B(z_Q, r)$ to $\partial\Omega$ at Q, and the radius r may be taken to be independent of the boundary point Q. One derives the desired estimates on the derivatives of f simply by restricting the function to any given B_Q. The estimates will of course be scaled by r but will be independent of Q.

The insights presented here were first recorded by E. M. Stein [**STE2**]. He realized that one could bootstrap these ideas to obtain explicit statements about nonisotropic smoothness of holomorphic functions. Let us introduce sufficient language that we may formulate his basic result.

Fix a domain $\Omega \subset\subset \mathbb{C}^n$ with C^2 boundary. Fix a tubular neighborhood U of $\partial\Omega$. This means that each $z \in U$ has a unique nearest point $\pi z \in \partial\Omega$. The point z lies along the normal line to $\partial\Omega$ through πz. The construction of U is an exercise with the implicit function theorem, or see [**MUN**],[**HIR**].

At each $z \in U$ we may perform a decomposition of space as follows. Associate to z the unit vector $\nu_z \equiv \nu_{\pi z}$ that is the outward normal to $\partial\Omega$ at πz. Let $\mathcal{N}_z$ be the complex line generated by ν_z. The orthogonal complement of $\mathcal{N}_z$, in the Hermitian inner product $\langle z, w\rangle \equiv z_1\bar{w}_1 + \cdots + z_n\bar{w}_n$, is denoted $\mathcal{T}_z$. We call $\mathcal{N}_z$ the *complex normal line* (or complex normal direction) at z and $\mathcal{T}_z$ the *complex tangent space* (or complex tangential directions) at z. Observe that $\mathcal{T}_z = \mathcal{T}_{\pi z}$ is canonical in the sense that it is the maximal complex subspace of the real tangent space $T_{\pi z}$ to $\partial\Omega$ at πz. However the choice of normal direction depends on the use of the Euclidean metric. It is worth noting, and the reader may check this for himself, that any other smooth Riemannian metric (other than the Euclidean metric) will give rise to a function theory that is equivalent to that which we are about to describe.

A C^1 curve $\gamma : [0,1] \to U \cap \Omega$ is called *normalized complex tangential* if

1: $\|\dot{\gamma}(t)\| \leq 1$ for all t;

2: $\dot{\gamma}(t) \in \mathcal{T}_{\gamma(t)}$ for each $t \in [0,1]$.

We write $\gamma \in \mathcal{C}^1$.

We shall now define function spaces on Ω that exhibit a certain global smoothness in *any* direction, but an increased smoothness along normalized complex tangential curves. For $0 < \alpha < 1$ recall that a (continuous) function f on Ω is said to be *Lipschitz* of order α, written $f \in \Lambda_\alpha$, if

$$|f(z+h) - f(z)| \leq C|h|^\alpha$$

for all $z, z+h \in \Omega$. To avoid technicalities we shall not discuss $\alpha \geq 1$, but refer the reader to [**KRA4**] for further details. Now we say that a continuous function g on Ω lies in $\Gamma_{\alpha,\beta}(\Omega), 0 < \alpha < \beta < 1$, if

1: $g \in \Lambda_\alpha(\Omega)$;

2: for any $\gamma \in \mathcal{C}^1$ it holds that $g \circ \gamma \in \Lambda_\beta[0,1]$, uniformly in γ.

Let us discuss this definition briefly. First, we are primarily interested in holomorphic functions. On any relatively compact subset of Ω, such a function is C^∞. Thus the only interesting smoothness phenomena take place at the boundary. That is why it is sufficient for us to mandate the extra smoothness (the β-order tangential smoothness) near the boundary only. The use of the curves $\gamma \in \mathcal{C}^1$ is rather clumsy, but it keeps the language fairly near the surface. It is more natural to express the nonisotropic smoothness of f using (weighted) mixed derivatives of the function, and making no reference to curves (see, for instance, [**STE2**],[**KRA4**]).

An even more natural language for expressing the non-isotropy is that of canonical metrics (again see [**KRA4**]). Alternatively, the structure of certain

Lie algebras provides a context for the nonisotropic analysis. These points of view will come out in the ensuing pages. For the present we shall strive for simplicity.

THEOREM 1.1.1 (STEIN). *Let* $\Omega \subset\subset \mathbb{C}^n$ *have* C^2 *boundary. Assume that* f *is a holomorphic function on* Ω *that lies in* $\Lambda_\alpha(\Omega), 0 < \alpha < 1/2$. *Then in fact* $f \in \Gamma_{\alpha,2\alpha}(\Omega)$. *[The converse is trivial.]*

A suitable version of this theorem is true for *any* real α, but we shall not discuss these extensions here. We shall give a quick rendition of the proof of the result as stated. (The reader may find full details in [**KRA1**], for instance.) Then we shall turn to a consideration of the geometry, which is the main point of this discussion.

PROOF OF THE THEOREM: It is enough to examine the proof when the domain is the ball $B \subseteq \mathbb{C}^2$. We retain the notation set down in the discussion of the ball preceding the statement of the theorem. We apply the Cauchy integral formula to f on a slightly shrunken version of $\mathbf{d}_n$ (of radius $.99\delta$, say). Thus, for $|z_1 - P_1| < .5\delta$, we have

$$f(z_1, 0) = \frac{1}{2\pi i}\oint_{|\zeta - P_1| = .99\delta} \frac{f(\zeta, 0)}{\zeta - z_1}\, d\zeta.$$

We differentiate in z_1 to obtain

$$\frac{\partial}{\partial z_1} f(z_1, 0) = \frac{1}{2\pi i}\oint \frac{f(\zeta, 0)}{(\zeta - z_1)^2}\, d\zeta. \tag{2}$$

Of course

$$\frac{1}{2\pi i}\oint \frac{1}{(\zeta - z_1)^2}\, d\zeta = 0$$

so we may rewrite (2) as

$$\frac{\partial}{\partial z_1} f(z_1, 0) = \frac{1}{2\pi i}\oint \frac{f(\zeta, 0) - f(P_1, 0)}{(\zeta - z_1)^2}\, d\zeta.$$

Now we set $z_1 = P_1$, exploit the Lipschitz condition to estimate the numerator of the integrand, and perform obvious estimates on the denominator to obtain

$$\left|\frac{\partial}{\partial z_1} f(P)\right| \leq C \cdot \delta^{\alpha - 1}. \tag{3}$$

An identical argument, using the disc $\mathbf{d}_t$, would allow us to prove that

$$\left|\frac{\partial}{\partial z_2} f(P)\right| \leq C \cdot \delta^{(\alpha - 1)/2}. \tag{4}$$

However (4) turns out not to be the best possible estimate, and is inadequate for our purposes.

Instead we apply our analysis to the holomorphic function $g \equiv \frac{\partial}{\partial z_1} f$, which (by the above arguments) satisfies a bound of the form $|g| \leq C\delta^{\alpha-1}$ at all points of the form $\{(P_1, \zeta) : |\zeta| \leq \sqrt{\delta}/2\}$. We then apply (C_2) to see that

$$\left| \frac{\partial^2}{\partial z_1 \partial z_2} f(P) \right| \leq C \cdot \delta^{\alpha-1-1/2} = C\delta^{\alpha-3/2}.$$

Now if $P = (P_1, 0) \in B$ and if (for convenience) P_1 is real and positive, then

$$\frac{\partial}{\partial z_2} f(z) = \int_0^{P_1} \frac{\partial^2}{\partial z_1 \partial z_2} f(t, 0)\, dt + \frac{\partial f}{\partial z_2}(0).$$

and we estimate this, using the last line, to have size not exceeding $C\delta^{\alpha-1/2}$.

Of course the estimates that we have just arrived are not dependent on P being of the special form $P = (P_1, 0)$ with P_1 real and positive. In fact it holds for any P near the boundary of B, of distance δ from ∂B, that a complex normal derivative of f has size not exceeding $\delta^{\alpha-1}$ and a complex tangential derivative has size not exceeding $\delta^{\alpha-1/2}$.

Armed with our separate estimates for the normal derivative of f and the tangential derivative of f, we proceed as follows. Let γ be a normalized complex tangential curve in B. If $h > 0$ is small then set $H = \gamma(h)$ and $Z = \gamma(0)$. Then

$$\begin{aligned} |f(\gamma(h)) - f(\gamma(0))| &\leq |f(H) - f(H - h^2 H/|H|)| \\ &\quad + |f(H - h^2 H/|H|) - f(Z - h^2 Z/|Z|)| \\ &\quad + |f(Z - h^2 Z/|Z|) - f(Z)| \\ &= I + II + III. \end{aligned}$$

Now I and III are estimated in just the same way: each difference is, by the Fundamental Theorem of Calculus, just an integral of $\partial f/\partial \nu$ over a segment of length h^2. Using our estimate on the normal derivative, we see that both I and III are bounded by $Ch^{2\alpha}$. If we define

$$\gamma^*(t) = \gamma(t) - h^2 \nu_{\gamma(t)}$$

then the difference in II is just $f(\gamma^*(h)) - f(\gamma^*(0))$. This is easily estimated using the Fundamental Theorem (the derivatives that fall on f are, by the Chain Rule, all tangential). The result is that $II \leq C \cdot h^{2\alpha}$.

The theorem is proved. ∎

The explicit dependence of the theorem on boundary geometry comes from the fact that a C^2 domain has, at each boundary point, an internally tangent ball of fixed size. However hindsight makes it clear that a phenomenon of complex analysis—especially the complex function theory of several variables—should not be dependent on a metric fact coming from *Euclidean geometry*. Put another way, we see from simple examples that Stein's theorem, as stated, cannot be optimal.

Consider for instance the domain

$$E_m = \{(z_1, z_2) \in \mathbb{C}^2 : |z_1|^2 + |z_2|^{2m} < 1\},$$

$m = 1, 2, \ldots$. If $P = (P_1, 0) \in E_m$ has distance $\delta > 0$ from the boundary then the largest tangential disc $\mathbf{d}_t$ with center at P will have radius at least $\delta^{1/(2m)}$. Thus the sort of estimates that we can prove (using Cauchy theory) for tangential derivatives at P of a holomorphic function must be *better* than the ones we obtained using the geometry of the ball. [In fact it may be worth noting that the paper [**KRA14**] explored how estimates of this nature, and the lack thereof, could be used to biholomorphically distinguish certain weakly pseudoconvex domains.]

Let us push the point further, and be even more severe in our criticism of the way that the theorem is proved. Why should we look at *rigid affine Euclidean discs* centered at P when we seek to estimate derivatives of f using Cauchy estimates on these discs? *Any analytic disc* $\phi : D \to \Omega$ will serve just as well. And it is conceivable, is it not, that the largest analytic disc through $P \in \Omega$, in a tangential or a normal direction, is larger than the obvious rigid affine Euclidean disc in that direction.

The natural language for treating the issues just raised is the metric of Kobayashi and Royden. This metric is defined as follows. Let $\Omega \subset\subset \mathbb{C}^n, P \in \Omega$, and let ξ be any element of $\mathbb{C}^n$. We think of (P, ξ) as an element of the (complexified) tangent bundle to Ω. Let $(\Omega, D)_{(P,\xi)}$ denote the collection of those holomorphic mappings ϕ of the unit disc $D \subseteq \mathbb{C}$ into Ω such that $\phi(0) = P$ and $\phi'(0)$ is a multiple of ξ. Let $\| \quad \|$ denote Euclidean length. Then set

$$F_K^\Omega(P, \xi) = \frac{\|\xi\|}{\sup_{\phi \in (\Omega, D)_{(P,\xi)}} \|\phi'(0)\|}.$$

This is the *infinitesimal Kobayashi/Royden metric* at the point P in the direction ξ.

The following universal property of the Kobayashi metric is immediate from the definition: Let $\Omega_1 \subseteq \mathbb{C}^m, \Omega_2 \subseteq \mathbb{C}^n$ be domains and let $\Phi : \Omega_1 \to \Omega_2$ be a holomorphic mapping. If (P, ξ) is an element of the tangent bundle to Ω_1 then

$$F_K^{\Omega_1}(P, \xi) \geq F_K^{\Omega_2}(\Phi(P), \Phi_*(P)\xi).$$

In particular, if Φ is biholomorphic then, by applying this observation both to Φ and to Φ^{-1}, we find that Φ *preserves* the Kobayashi/Royden metric.

Some examples will serve to make the Kobayashi/Royden metric more tangible.

EXAMPLE 1.1.2. Let $\Omega \subseteq \mathbb{C}^2$ be the unit ball. Let P be the origin and, for convenience, take $\xi = (1, 0)$. The mapping $\phi \in (\Omega, D)_{(P,\xi)}$ given by $\phi(\zeta) = (\zeta, 0)$ shows that $F_K^\Omega(P, \xi) \leq 1$. On the other hand, if ψ is any element of $(\Omega, D)_{(P,\xi)}$ and if π_1 is projection on the first variable then we may apply the Schwarz lemma to the classical holomorphic function $\pi_1 \circ \psi$ to see that $|\psi'(0)| \leq 1$. It follows that $F_K^\Omega(P,\xi) \geq 1$, hence $F_K^\Omega(P,\xi) = 1$.

If now Q is *any* point of the ball $\Omega = B$ and η is a direction at Q then there is a biholomorphic self-mapping Ψ of the ball that maps the origin to Q and whose differential maps $\xi = (1,0)$ to a multiple of η (see [**RUD1**] for a detailed treatment of the biholomorphic self-maps of the ball). Thus we may use the invariance properties of the Kobayashi metric to determine, from the calculation in the previous paragraph, that

$$F_K^\Omega(Q,\eta) = \sqrt{\frac{\|\eta\|^2}{1-|Q|^2} + \frac{|\langle Q,\eta\rangle|^2}{(1-|Q|^2)^2}}$$

(see [**GRA1**]). We may read off from this formula that, if $Q = (1-\delta, 0)$ and $\eta = (0,1)$ then $F_K^\Omega(Q,\eta) = 1/\sqrt{2\delta - \delta^2}$. In particular, the mapping $\phi(\zeta) = (1-\delta, \sqrt{2\delta-\delta^2}\zeta)$ is extremal for the Kobayashi/Royden metric in the tangent direction at Q. If instead, for this choice of Q, we take $\eta = (1,0)$, then the formula tells us that $F_K^\Omega(P,\eta) = 1/(2\delta - \delta^2)$ and the mapping

$$\mu(\zeta) = \left(\frac{\zeta + (1-\delta)}{1+(1-\delta)\zeta}, 0\right)$$

is extremal for the Kobayashi/Royden metric in the normal direction at P.

In summary we find, for the domain $\Omega \subseteq \mathbb{C}^2$ the unit ball, that the use of rigid discs to measure "complex extent" in any given direction is no different from using analytic discs (although the exact extremal disc in the normal direction is not the "obvious" extremal disc from (C_1)). Thus the proof of Stein's theorem is tight in this case. Moreover, the example $f(z_1, z_2) = (z_1 - 1)^\alpha$ shows that the statement of Stein's theorem is sharp when the domain is the ball.

In detail, it is plain that this function f is Λ_α, and no better, in normal directions. Now let $P = (1-\delta, 0)$ and $Q = (1-2\delta, \sqrt{\delta})$ be points of the ball. We leave it as an exercise to check that these points may be joined by a normalized complex tangential curve of length $\approx \sqrt{\delta}$. Then $|f(P) - f(Q)| \approx \delta^\alpha \approx |P-Q|^{2\alpha}$. So we see that $f \in \Gamma_{\alpha,2\alpha}$ but no better. ∎

EXAMPLE 1.1.3. Let $\Omega = E_m$, the complex ellipsoid or "egg" that was discussed earlier in this section. If $P = (1-\delta, 0)$ and $\xi = (0,1)$ then the map $\zeta \mapsto (1-\delta, \delta^{1/2m}\zeta)$ shows that $F_K^\Omega(P,\xi) \leq \delta^{-1/2m}$. Also, for $\eta = (1,0)$, it holds that $F_K^\Omega(P,\eta) \approx \delta^{-1}$. A calculation (see [**BFKKMP**] for the exact result) shows that, up to a constant of comparison, the first estimate is sharp. The second estimate follows as in the case of the ball. These results are not surprising because the egg domain covers the ball via the mapping $(z_1, z_2) \mapsto (z_1, (z_2)^m)$ thus there should be some (local) correspondence between extremal discs. In fact the proof in [**BFKKMP**] consists of a careful exploitation of this covering map.

Thus we see in the example of the egg domain that, on the one hand, the geometry of (internally tangent) *balls* is inadequate to study "complex extent."

Moreover the example $f(z_1, z_2) = (z_1 - 1)^\alpha$ shows that the smoothness in complex tangential directions increases to $2m \cdot \alpha$. On the other hand, the discussion in the preceding paragraph demonstrates that rigid affine discs still capture the essence of the situation. So nothing new appears in the present example. ■

With the next two examples we will provide the *raison d'etre* for the Kobayashi metric point of view—in particular we will demonstrate that the use of complex analytic discs, as opposed to rigid affine discs, reveals much more information.

Notice that in the preceding two examples it was the case that, for P near $\partial\Omega$ and ξ pointing in the complex normal direction, it holds that $F_K^\Omega(P, \xi) \approx \|\xi\|/\delta_\Omega(P)$. [Here $\delta_\Omega(P)$ is the Euclidean distance of P to $\partial\Omega$.] Indeed, for reasonable pseudoconvex domains, such as strongly pseudoconvex domains in $\mathbb{C}^n$ and finite type domains in $\mathbb{C}^2$, this sort of estimate is known to always hold (see [**GRA1**], [**CAT4**]). [On the other hand, it is *not* universal, as we shall see in a moment.] In general, if Ω is any smoothly bounded domain in $\mathbb{C}^n$ and if P is a point of Ω that is very near the boundary then we may apply the distance decreasing property of the Kobayashi metric to the inclusion map of the internally tangent ball to $\partial\Omega$ at πP to see that $F_K^\Omega(P, \nu_P) \leq C \cdot \delta^{-1}$. The reverse inequality is always delicate to prove, and is in general false as the following considerations show.

There are two types of examples that we wish to consider.

EXAMPLE 1.1.4. We exhibit a smooth, pseudoconvex domain Ω for which the Kobayashi metric does not satisfy

$$F_K^\Omega(P, \xi) \geq C \cdot \delta_\Omega(P)$$

on unit vectors ξ that are essentially normal and points P that are tending to the boundary in the normal direction (this example is from [**FOK**]). This is not quite enough to show that the domain is incomplete in the Kobayashi metric, but serves to (essentially) disappoint the conjecture that $F_K(z, \nu) \sim \delta(z)^{-1}$. The details follow:

LEMMA 1.1.5. *There is a C^∞ subharmonic function ρ on $\mathbb{C}$ with the following properties:*

(i) ρ vanishes to infinite order at 0;

(ii) There is a sequence $r_n \searrow 0$ rapidly and a sequence $\delta_n \searrow 0$ such that, for each n,

$$\rho(z) < \delta_n + n\frac{\delta_n}{r_n}\mathrm{Re}\, z \qquad \text{if} \quad |z| \leq r_n.$$

Suppose for the moment that the lemma has been proved. Set

$$\Omega = \{(z, w) \in \mathbb{C}^2 : \mathrm{Re}\, w + \rho(z) < 0\} \cap B(0, R),$$

where $B(0, R)$ is a large ball. Set $p_n = (0, -2\delta_n)$ and

$$X_n = \frac{\partial}{\partial z} - n\frac{\delta_n}{r_n}\frac{\partial}{\partial w}.$$

The normal component of X_n is plainly equal to $n\delta_n/r_n$. For the purposes of this example, we may think of X_n as a 'normal' vector.

The "naive", or expected value of the infinitesimal Kobayashi metric at p_n in the direction X_n is

$$\frac{n\delta_n/r_n}{\delta_n}$$

(because the point p_n is distance δ_n from the boundary and the vector has Euclidean length $n\delta_n/r_n$—compare with the strongly pseudoconvex case as in [**GRA1**]). The actual infinitesimal Kobayashi metric is not greater than a constant times $1/r_n$ since the analytic disc

$$\mathbf{d}_n \equiv \{(z,w) : w = -2\delta_n - n(\delta_n/r_n)z, |z| \leq r_n\}$$

lies entirely in Ω. So we see that

$$F_K(p_n, X_n) \leq \frac{1}{r_n},$$

in particular that

$$\frac{\delta_n F_K(p_n, X_n)}{n\delta_n/r_n} \to 0.$$

So $F_K(p_n, X_n)$ does not blow up like $1/\delta_n$.

It remains to verify the lemma. We sketch the proof. For $C_n > 0$ large we define

$$R_n(\zeta) = \begin{cases} \left(\frac{1}{2} + \operatorname{Re}\zeta + \frac{\log|\zeta|}{C_n}\right) \vee 0 & \text{if } \operatorname{Re}\zeta > -\frac{1}{4} \\ \frac{1}{2} + \operatorname{Re}\zeta + \frac{\log|\zeta|}{C_n} & \text{if } \operatorname{Re}\zeta \leq -\frac{1}{4}. \end{cases}$$

Observe that $R_n(\zeta) \equiv 0$ for ζ near 0 and $R_n(\zeta) < 1 + \operatorname{Re}\zeta$ if $|\zeta| \leq n$.

For each n we scale R_n and multiply the resulting function by a suitable constant to obtain a function ρ_n so that $\rho_n \equiv 0$ on a disc $\Delta_n = \{|\zeta| < s_n\}$ with $s_n \searrow 0$. If $r_n < s_{n-1}$ then it can be arranged that

$$\rho_n(\zeta) < \delta_n + n\frac{\delta_n}{r_n}\operatorname{Re}\zeta \qquad \text{if} \quad |\zeta| \leq r_n.$$

Finally, we set $\rho(\zeta) = \sum_n \rho_n(\zeta)$. One checks that ρ satisfies all the required properties.

In summary, we have constructed an example of a smooth pseudoconvex domain for which the infinitesimal Kobayashi metric does not blow up, in (essentially) the normal direction, like the reciprocal of the distance to the boundary.

To obtain an even more dramatic example where the "one over distance" estimate in the normal direction does not obtain, it is convenient to study a non-pseudoconvex domain.

EXAMPLE 1.1.6. Consider the domain

$$\Omega = \{z \in \mathbb{C}^2 : 1 < \|z\| < 4\}. \tag{5}$$

Let $P_\delta = (-1-\delta, 0) \in \Omega$. For $\delta > 0$ small, we let $\xi = \nu_{P_\delta} = (1,0)$ be the normal vector to $\partial\Omega$ corresponding to P_δ. We shall show that $F_K^\Omega(P_\delta, \xi) \approx \delta^{-3/4}$.

The proof proceeds in two steps. First, we exhibit a mapping $\phi \in (\Omega, D)_{(P_\delta,\xi)}$ such that $\phi(0) = P_\delta, \phi'(0) = c\xi$, and $c \approx \delta^{3/4}$. Second, we show that *for all* $\psi \in (\Omega, D)_{P_\delta,\xi}$ it holds that $|\psi'(0)| \le C \cdot \delta^{3/4}$.

For the first step, set

$$\phi(\zeta) = (-1 - \delta + (\delta^{3/4}/10)\zeta, \zeta^2).$$

Observe that $\phi(0) = P_\delta$ and $\phi'(0) = (\delta^{3/4}/10, 0)$. If we can show that the image of ϕ, acting on the unit disc, lies in Ω then we will know that

$$F_K^\Omega(P_\delta, \xi) \le C \cdot \delta^{-3/4}.$$

Now the estimate $|\phi(\zeta)| < 4$ is trivial. On the other hand, we see that

$$\|\phi(\zeta)\| > 1$$

if and only if

$$2\delta + \delta^2 + \frac{1}{100}\delta^{3/2}|\zeta|^2 - \frac{1}{5}(1+\delta)\delta^{3/4}\mathrm{Re}\,\zeta + |\zeta|^4 > 0. \tag{6}$$

There are now two cases to consider

a: If $|\zeta| < 5\delta^{1/4}$ then

$$\left|\frac{1}{5}(1+\delta)\delta^{3/4}\mathrm{Re}\,\zeta\right| < (1+\delta)\delta^{3/4}\delta^{1/4} < 2\delta.$$

Thus (6) holds and $\phi(\zeta) \in \Omega$ for these values of ζ, as desired.

b: If $|\zeta| \ge 5\delta^{1/4}$ then

$$\left|\frac{1}{5}(1+\delta)\delta^{3/4}\mathrm{Re}\,\zeta\right| \le \left|\frac{1}{5}(1+\delta)(|\zeta|/5)^3\mathrm{Re}\,\zeta\right|.$$

This, in turn, is majorized by $|\zeta|^4$. Therefore (6) holds and $\phi(\zeta) \in \Omega$ for these values of ζ, as desired.

Thus the image of ϕ lies in Ω and we have established the inequality in one direction.

For the inequality in the other direction, we must consider all mappings $\phi : D \to \Omega$ such that $\phi(0) = P_\delta$ and $\phi'(0)$ is a multiple of $\xi = (1,0)$ and we must establish that $|\phi'(0)|$ is majorized by $C \cdot \delta^{3/4}$, with C independent of ϕ.

If $\delta > 0$ is small, then we let

$$\mathcal{R}_\delta \equiv \{w \in \mathbb{C} : 1 - \delta < |w| < 4\}.$$

Now we claim that if $Q \in \mathcal{R}_\delta$ and $\xi \in \mathbb{C}$ is any Euclidean unit vector then

$$F_K^{\mathcal{R}_\delta}(Q, \xi) \approx \mathrm{dist}\,(Q, \partial\Omega)^{-1}. \tag{7}$$

The constants of comparison here will be *independent of* δ. This is an interesting point, and not explained elsewhere in the literature; we will prove it at the end of the section.

Now let $\phi : D \to \Omega$ with $\phi(0) = P_\delta$ and $\phi'(0)$ a multiple of ξ. We write $\phi(\zeta) = (\phi_1(\zeta), \phi_2(\zeta))$. An easy calculation shows that $|\phi_2(\zeta)| \le 4|\zeta|^2$. If $|\zeta| \le \delta^{1/4}/2$ then

$$|\phi_2(\zeta)| \le 4\delta^{1/2}/4 = \delta^{1/2}.$$

Therefore, for such ζ,

$$|\phi_1(\zeta)| \ge \sqrt{1 - |\phi_2(\zeta)|^2} \ge \sqrt{1-\delta} > 1 - \delta.$$

Thus the function $g(\zeta) \equiv \phi_1(\delta^{1/4}\zeta/2)$ maps the disc D to $\mathcal{R}_\delta$ with $g(0) = -1 - \delta$. By our estimates on the Kobayashi metric for the domains $\mathcal{R}_\delta$, we conclude that

$$\frac{\delta^{1/4}}{2}|\phi_1'(0)| = |g'(0)| \le C \cdot \delta.$$

As a result,

$$|\phi_1'(0)| \le C' \cdot \delta^{3/4}.$$

Since ϕ was arbitrary, we have determined that

$$F_K^\Omega(P_\delta, \xi) \ge C'' \cdot \delta^{-3/4}.$$

In summary, we have established that, for the particular domain Ω defined in line (5), the Kobayashi metric satisfies

$$F_K^\Omega(P_\delta, \xi) \approx \delta^{-3/4}.$$

Similar calculations (see [**KRA3**]) show that, for λ in the range $3/4 \le \lambda < 1$ the domain

$$\Omega_\lambda \equiv \{(z_1, z_2) \in \mathbb{C}^2 : 1 < |z_1|^2 + |z_2|^{1/(2-2\lambda)} < 4\}$$

satisfies

$$F_K^{\Omega_\lambda}(P_\delta, \nu) \approx \delta^{-\lambda}. \tag{8}$$

Here $P_\delta = (-1-\delta, 0)$, δ small, and $\nu = (1,0)$ is the normal vector at P_δ. The domain Ω_δ has C^2 boundary. *Any* strongly pseudoconvex domain U satisfies an estimate like (8) with $\lambda = 1$. Thus we have discovered an important piece of geometric/analytic information. If U is *any* smoothly (C^2) bounded domain in $\mathbb{C}^n$, if $P \in \Omega$ is a point near the boundary and δ its Euclidean distance to the boundary then

$$C \cdot \delta^{-3/4} \le F_K^\Omega(P, \nu) \le C' \cdot \delta^{-1}.$$

To see this, first note that there is an $r' > 0$ such that there is an externally tangent Euclidean ball $B'_P = B(z'_P, r')$ of uniform radius r' at each point $P \in \partial U$. On the other hand, the domain U has some finite diameter D. We may assume that $r' << D$. Then the spherical shell $S_P \equiv B(z'_P, 2D) \setminus \bar{B}'_P$ *contains* U. See Figure 1.2.

If z is a point situated in U, along the interior unit normal to ∂U at P, and at distance $\delta > 0$ (small) from P, then

$$F_K^U(z, \nu_z) \ge F_K^{S_P}(z, \nu_z) \approx C \cdot \delta^{-3/4}.$$

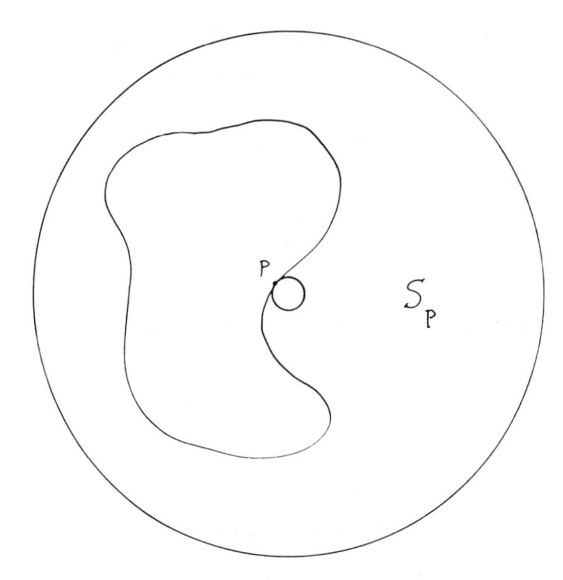

FIGURE 1.2

Thus we see that, for smoothly bounded domains, the $\delta^{-3/4}$ estimate is the worst possible for the Kobayashi metric. On the other hand, comparison with an internally tangent ball B_P at P (see the earlier discussion in this section) shows that $F_K^U(z, \nu_z) \leq C \cdot \delta^{-1}$. See Figure 1.3. The calculations in [**KRA3**] show that all values of the exponent between -1 and $-3/4$ can actually occur for smoothly (C^2) bounded domains.

On a strongly pseudoconvex domain, or on a domain of finite type in $\mathbb{C}^2$ (see Section 1.3), it is known that the Kobayashi metric blows up in the normal direction like (1/distance). Estimates on the Kobayashi metric in *all directions* can be used to characterize finite type, at least in dimension 2 (see [**FU, YU1**]). The trouble that arises in higher dimensions is that the correct way to detect type is to use possibly singular varieties. Then the obvious idea of translating the variety of best contact into the domain (see Figure 1.4) to obtain a competitor for the Kobayashi/Royden disc fails, because the derivative of the parameterizing map vanishes. J. Yu [**YU4**] has recently developed a variant of the Kobayashi metric that is based on higher derivatives of analytic discs and which can be used to characterize points of finite type in any dimension. Higher order variants of the Kobayashi metric also appear in [**WU2**], where they are used to study hyperbolicity.

The estimate for the Kobayashi metric on $\Omega_{3/4}$ forces a re-evaluation of Stein's theorem in this case. For a proof similar to that of Stein's theorem, but using the extremal disc constructed for the Kobayashi metric estimate in Example 1.6 (rather than the naive rigid affine disc of radius δ at a point having distance δ from the boundary) shows that if f is holomorphic on Ω and $f \in \Lambda_\alpha$ then f continues C^∞ to the inner boundary of the shell-shaped region.

In detail, assume for simplicity that $0 < \alpha < 1/2$. Let $\gamma : [0, h] \to \Omega$ be a

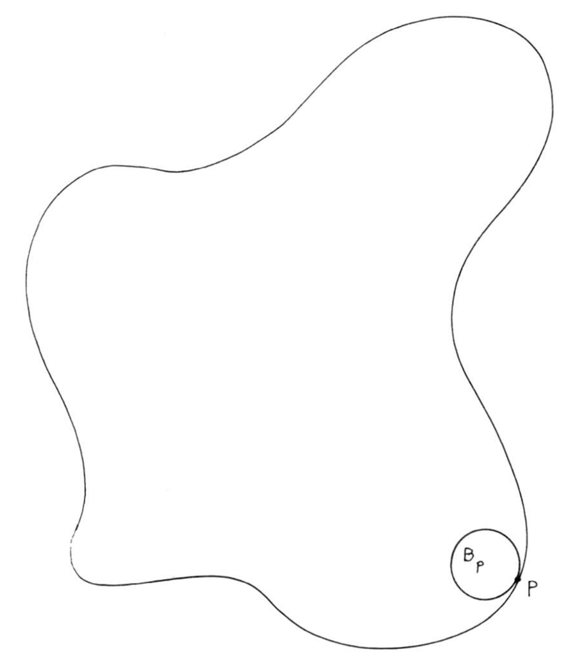

FIGURE 1.3

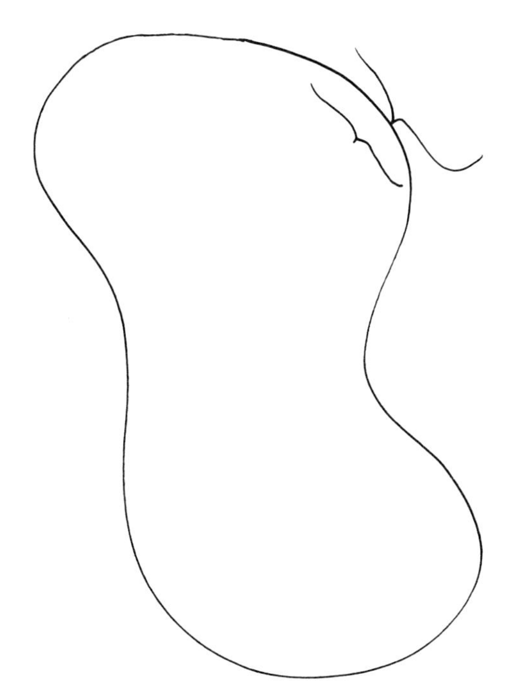

FIGURE 1.4

parametrization by arc length of a segment normal to $\partial\Omega$. We estimate $|f(\gamma(h))-f(\gamma(0))|$ using the fundamental theorem of calculus (as in the proof of Stein's theorem above). That is, this difference is expressed as an integral of $(d/dt)f\circ\gamma$. But that derivative satisfies an estimate of size $\delta^{-3/4}$. When this is integrated, along a segment of length $|h|$, the result is a gain of $1/4$ in the exponent.

The result is that the function f is in $\Lambda_{\alpha+1/4}$ along the standard normal direction to the boundary. By a theorem of [**KRA11**], we may conclude that $f\in\Lambda_{\alpha+1/4}$. It is critical to notice that the amount of improvement is *independent of* α. In fact arguments using higher order differences may be constructed to obtain an improvement of order $1/4$ no matter what the initial size of α. Thus, by iteration, an f that begins in any Lipschitz class ends up being in C^∞ up to the boundary, as desired.

What is the meaning of this result? In point of fact the Hartogs extension phenomenon tells us that any holomorphic function on $\Omega_{3/4}$ continues analytically to $B(0,1)$. What we have done, using the Kobayashi metric and extremal discs instead of rigid affine discs, is to rediscover the Hartogs extension phenomenon in the category of Lipschitz spaces. The argument we have given adapts to any boundary point at which the Levi form has a negative eigenvalue. It tells us that, near such a boundary point, Lipschitz holomorphic functions are automatically C^∞ up to the boundary. This is just the Hartogs theorem in a (Lipschitz) disguise. Details of these ideas may be found in [**KRA4**].

Let us conclude the section by proving the claim about the blowup of the Kobayashi metric on a smoothly bounded domain in the plane (see [**KRA9**] for related ideas). We begin with a formal statement:

LEMMA 1.1.7. *Let $\Omega\subseteq\mathbb{C}$ be a bounded domain with $C^{1+\epsilon}$ boundary. There is a constant $C>0$ such that, for any $\zeta\in\Omega$ and any direction ξ it holds that*

$$\frac{1}{C}\frac{|\xi|}{\delta_\Omega(\zeta)}\le F_K^\Omega(\zeta,\xi)\le C\frac{|\xi|}{\delta_\Omega(\zeta)}.$$

The constant C depends only on $\|\rho\|_{C^{1+\epsilon}}$, where ρ is any defining function for $\Omega:\Omega=\{\zeta\in\mathbb{C}:\rho(\zeta)<0\},\nabla\rho\neq 0$ on $\partial\Omega$.

PROOF. On any compact subset of Ω, the Kobayashi metric is comparable to the Euclidean metric. So the assertion is of interest only near the boundary.

We may choose a number $R>0$ such that for any $P\in\partial\Omega$ the set $D(P,R)\cap\Omega$ (with $D(P,R)$ a disc of center P and radius R) is topologically trivial. By the Riemann mapping theorem, there is a conformal mapping

$$\phi_P:D(P,R)\to D(0,1).$$

Moreover, standard regularity results for conformal mappings (see [**BEK**] or [**KRA7**]) show that the mapping ϕ is C^1 up to (the interior of) that portion of the boundary of $D(P,R)\cap\Omega$ that comes from Ω. The estimate (7) that we seek holds *prima facie* on the disc because in that venue we have an explicit formula

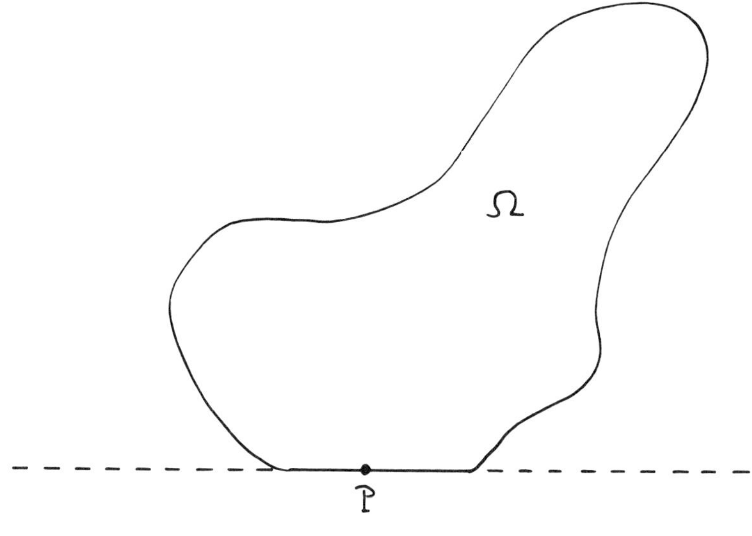

FIGURE 1.5

for the metric. By the invariance of the metric under conformal mappings, and the C^1 continuability of the mapping to the boundary, we see that the desired estimate also holds on $D(P,R) \cap \Omega$ *near points of* $\partial\Omega$.

Thus it remains to check that $F_K^{\Omega}(P,R)$ and $F_K^{D(P,R)\cap\Omega}$ are comparable at points near to $\partial\Omega$. Fix $P \in \partial\Omega$. After a conformal mapping, we may suppose that $\partial\Omega$ is flat near P and the picture is as in Figure 1.5.

Shrinking R if necessary, we may suppose then that $D(P,R)\cap\Omega$ is just a half disc. Notice (Figures 1.6 and 1.7) that $D(P,R) \cap \Omega \subseteq \Omega \subseteq W$.

Then obvious comparisons (see details on the Kobayashi metric in the Section 1.4) show that

$$F_K^W \leq F_K^{\Omega} \leq F_K^{D(P,R)\cap\Omega}.$$

The first and last expressions in this displayed material are comparable by conformal mapping. Both of these are comparable to $1/\delta_\Omega$ because we can explicitly map a half disc to a disc with a conformal map that is smooth across the smooth parts of the boundary. That completes the proof.

Notice that the constants of comparability depend on the boundary smoothness of the relevant conformal mappings, and this in turn depends only on the quantity $\|\rho\|_{C^1}$. ∎

It is worth mentioning that at least some boundary smoothness is required in order for the lemma to be true. For instance, the Kobayashi metric on $D \setminus \{0\}$ does not satisfy the $1/(\text{distance})$ estimate: the universal covering map shows that this is the case. On the other hand (as was pointed out to me by Joel Shapiro), a domain with Jordan boundary curves satisfies the "one over distance" estimate, as can be seen by examining the uniformization mapping.

I do not know the sharp condition that guarantees the $1/\delta$ estimate. However

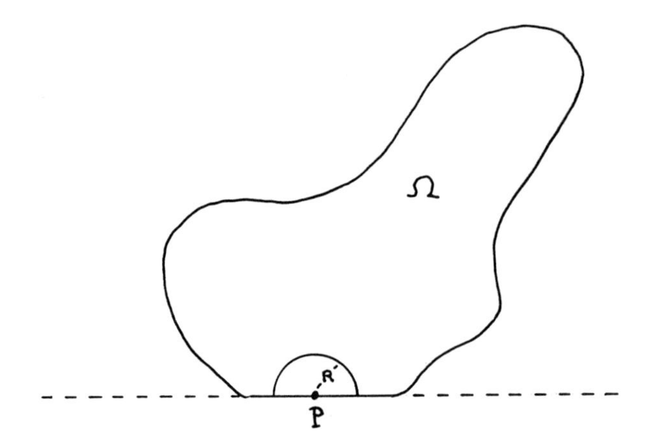

FIGURE 1.6

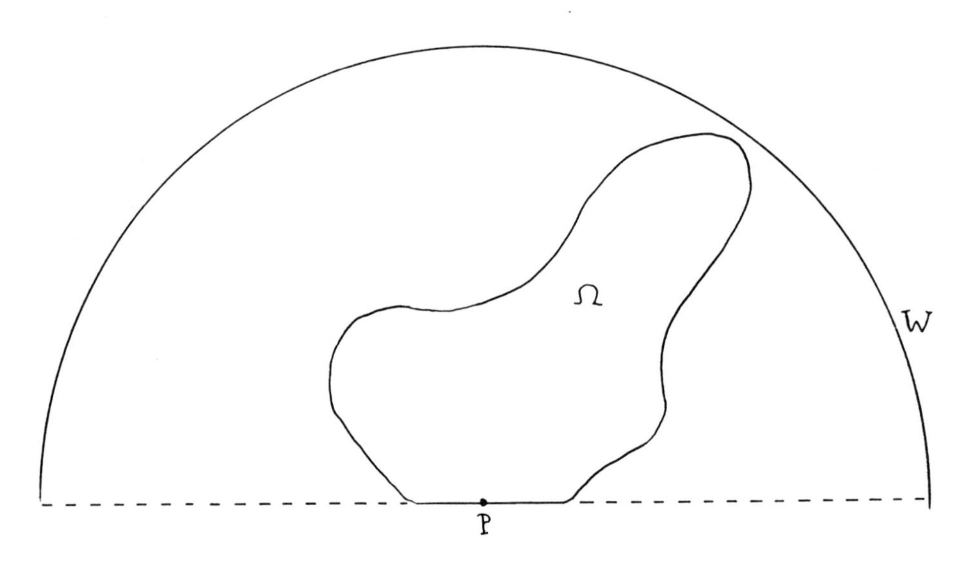

FIGURE 1.7

note the following: if $P \in \partial\Omega$ and if there is a *peaking function* ϕ_P for the point P (i.e. a function ϕ_P that is continuous on $\bar{\Omega}$, holomorphic on Ω, satisfies $\phi_P(P) = 1$ and $|\phi_P(z)| < 1$ for $z \neq P$) then Hopf's lemma shows that $(\partial/\partial\nu)\text{Re}\,\phi_P(P) > 0$. It follows from elementary estimates (or use comparisons with the Carathéodory metric as in [**KRA9**]) that the Kobayashi metric satisfies the "one over distance estimate" along the normal to P. Since peaking functions are not difficult to construct once one can solve the Dirichlet problem (construct the logarithm of the real part—see [**GAM**]), these considerations suggest to me that the "one over distance estimate" near a boundary point P is closely connected to whether the point P has a *barrier* (see [**AHL1**]).

1.2. From Classical to Modern Analysis. The history of analysis in the twentieth century has an interesting ontology. In the early days, most work was on the circle. This is a one dimensional compact group—every point is the same as every other—and working in this environment allows one to be unencumbered by topology or geometry. Moreover, one can and should pass to the disc, and harmonic and analytic function theory, for extra tools and ideas. Passing to the real line $\mathbb{R}$ involves dealing with a non-compact space and certain convergence problems at infinity (see Hoffman's book [**HOF**] for an elegant treatment), but nothing fundamentally new comes up.

Analysis in $\mathbb{R}^N, N \geq 2$, offers many new features. First, there is no longer just one group in sight; there are instead three. These are the translations, dilations, and rotations. Stein [**STE1**] makes a forceful case for how these groups shape the analysis of Euclidean space. Singular integral operators, particularly the N-tuple $(T_1, \ldots, T_N)$ of Riesz transforms, with $T_j\phi \equiv K_j * \phi$ and $K_j \equiv x_j/|x|^{N+1}$, are seen to be canonical when viewed in terms of their invariance under these groups. So are the fractional integral operators

$$\phi \mapsto |x|^{\alpha-N} \quad , \qquad 0 < \alpha < N,$$

of Riesz.

Also in dimensions two and higher the open sets no longer have a canonical form. Whereas in one dimension they are just intervals and their pairwise disjoint unions, in higher dimensions there is no analogous fact. A good substitute is the Whitney decomposition of an open set (see [**STE1**]).

The construction breaks an open set into boxes with disjoint interiors, each of which has distance to the boundary comparable to its diameter. Refer to Figure 1.8.

Whitney originally devised this decomposition in order to prove his theorem about continuing the germ of a smooth function from an arbitrary closed set to all of space. It is an interesting exercise to use the Whitney decomposition to prove that *any closed set* in $\mathbb{R}^N$ is the zero set of a C^∞ function. (This is in sharp contrast to the structure theorem for zero sets of real analytic functions—see [**KRP1**],[**KRP2**]).

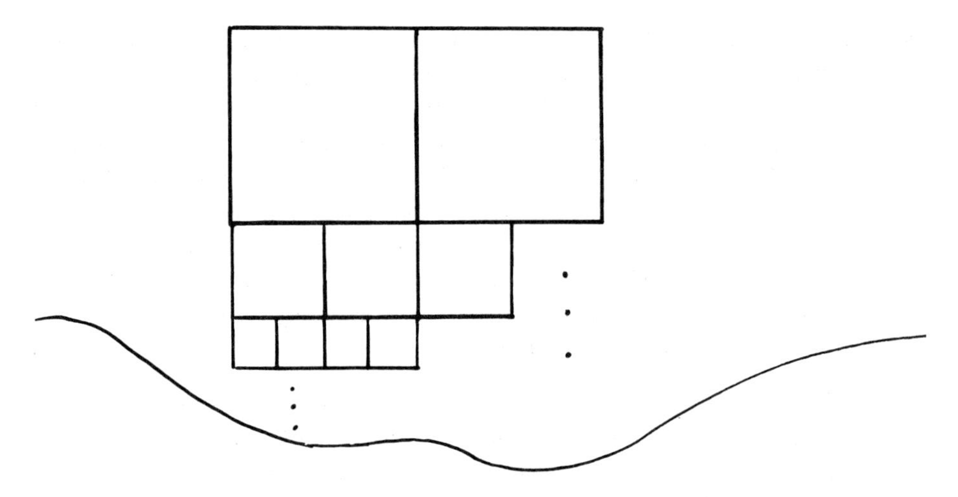

FIGURE 1.8

The next step, which really only got under way in the 1970's, was to study nonisotropic analysis. As indicated in Section 1, such considerations already arise naturally in studying the heat equation on the upper half space. However the analysis for the heat equation is still fundamentally Euclidean in nature: the fundamental open sets are boxes with sides parallel to the axes (of dimensions $\delta \times \delta^2$) and the situation is essentially abelian. We shall develop this idea, of how a certain geometry fits a certain kernel, as this section and the next develop.

The ideas begin to fall into place as we examine a situation which is close to, but just a bit more complex than, that which is familiar. We begin with the analysis of the Heisenberg group.

The Heisenberg group acts naturally on the boundary of the unit ball in $\mathbb{C}^n$ in a simply transitive fashion. By this we mean the following: Let $n \geq 2$ and let $B \subseteq \mathbb{C}^n$ be the unit ball. Let $\mathcal{U} = \{w \in \mathbb{C}^n : \operatorname{Im} w_1 > \sum_{j=2}^n |w_j|^2\}$. Define $\Phi(z) = (w_1, \ldots, w_n)$, where

$$\begin{aligned} w_1 &= i \cdot \frac{1 - z_1}{1 + z_1} \\ w_j &= \frac{z_j}{1 + z_1}, \quad j = 2, \ldots, n. \end{aligned}$$

One may check that Φ maps B biholomorphically onto $\mathcal{U}$. The domain $\mathcal{U}$ is the standard unbounded realization of the bounded symmetric domain B. It is called a *Siegel upper half-space* of type II (see [**KAN**]).

The set $\mathbb{C}^{n-1} \times \mathbb{R}$ can be equipped with the multiplicative structure

$$(\zeta, t) \cdot (\xi, s) = (\zeta + \xi, t + s + 2 \operatorname{Im} \zeta \cdot \bar{\xi}),$$

where $\zeta\cdot\bar{\xi} \equiv \sum_{j=1}^n \zeta_j\bar{\xi}_j$. This binary operation makes $\mathbb{C}^{n-1}\times\mathbb{R}$ into a non-abelian group (called the *Heisenberg group* and denoted by $\mathbb{H}_{n-1}$).

If $(t+i|z'|^2, z_2, \ldots, z_n) \in \partial\mathcal{U}$, where $z' = (z_2, \ldots, z_n)$, then identify this point with $(z', t) \in \mathbb{H}_{n-1}$. Thus $\partial\mathcal{U}$ is a group in a natural way. If $w \in \mathcal{U}$, then we write $\rho(w) = \operatorname{Im} w_1 - |w'|^2 > 0$ and

$$\begin{aligned} w &= ((\operatorname{Re} w_1 + i|w'|^2) + i(\operatorname{Im} w_1 - |w'|^2), w_2, \ldots, w_n) \\ &\equiv (\tilde{w}_1, \ldots, \tilde{w}_n) + (i\rho(w), 0, \ldots, 0). \end{aligned}$$

If $(z', t) = g \in \mathbb{H}_{n-1}$, then let the action of g on $\mathcal{U}$ be given by

$$gw = g(\tilde{w}_1, \ldots, \tilde{w}_n) + (i\rho(w), 0, \ldots, 0),$$

where $g(\tilde{w}_1, \tilde{w}_2, \ldots, \tilde{w}_n)$ denotes the standard group operation in $\mathbb{H}_{n-1}$. The reader should check that in fact this is a *holomorphic* mapping of $\mathcal{U}$. Indeed, each $g \in \mathbb{H}_{n-1}$ induces an element of $\operatorname{Aut}\mathcal{U}$ (the biholomorphic self-maps of $\mathcal{U}$). Obviously $\mathbb{H}_{n-1}$ acts simply transitively on $\partial\mathcal{U}$, just as a group acting on itself. The group acts on the full space $\mathcal{U}$ as though $\mathcal{U}$ were the semi-direct product of $\mathbb{H}_{n-1}$ with $\mathbb{R}^+$ (see [**CRG**]).

Now fix attention on $\mathbb{H}_1$. There are three distinguished one parameter subgroups of Aut (Ω): $\alpha(\tau) = (\tau + i0, 0)$, $\beta(\tau) = (0 + i\tau, 0)$, and $\gamma(\tau) = (0 + i0, -2\tau)$. [The number -4 appears so that the commutators below come out nicely.] Each of these subgroups corresponds to a vector in the Lie algebra for $\mathbb{H}_1$. More precisely, if $f \in C_c^\infty(\mathbb{H}_1)$ and $g = (z, s) \in \mathbb{H}_1$ then

$$Xf(g) = \frac{d}{d\tau} f(g\cdot\alpha(\tau))\bigg|_{\tau=0} = \frac{d}{d\tau} f\big(z+t, s+\operatorname{Im}(z\cdot t)\big)\bigg|_{\tau=0} = \frac{\partial}{\partial x} + y\frac{\partial}{\partial t} f(g).$$

$$Yf(g) = \frac{d}{d\tau} f(g\cdot\beta(\tau))\bigg|_{\tau=0} = \frac{d}{d\tau} f\big(z+it, s+\operatorname{Im}(z\cdot\overline{it})\big)\bigg|_{\tau=0} = \frac{\partial}{\partial y} - x\frac{\partial}{\partial t} f(g).$$

$$Tf(g) = \frac{d}{d\tau} f(g\cdot\gamma(\tau))\bigg|_{\tau=0} = \frac{d}{d\tau} f\big(z, s-2-2\tau\big)\bigg|_{\tau=0} = \frac{\partial}{\partial t} f(g).$$

In the setting corresponding to two complex variables, the Lie algebra has three generators X, Y, T that satisfy the commutation relation

$$[X, Y] = XY - YX = T$$

and all other commutators are zero (*Exercise:* work this out). It results that the natural dilation structure in the Lie algebra is

$$aX + bY + cT \mapsto raX + raY + r^2cT,$$

for T should be thought of as a second derivative in X and Y. The (globally defined) exponential map pushes this dilation structure downstairs to the group. The non-isotropy of the situation forces one to consider balls which, when centered at the origin, have radius r^2 in the T direction and radii r in the X, Y directions.

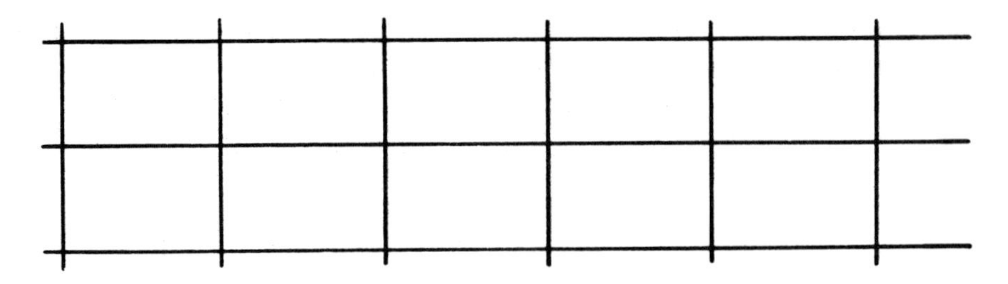

FIGURE 1.9

The important thing to keep in mind about the Heisenberg group is that the balls are *not* laid out like tiles (Figure 1.9).

In fact, as the center of a ball of fixed radius r moves away from the origin the ball *twists* (Figure 1.10). A calculation shows exactly what is going on:

Consider the simplest Heisenberg group $\mathbb{H}_1$. Topologically the group is $\mathbb{C} \times \mathbb{R}$. The group law is

$$(z,t) \cdot (z',t') = \big(z + z', t + t' + 2\mathrm{Im}\,(z\bar{z}')\big).$$

Consider the ball centered at the origin that is given by

$$B = \{(z,t) : |z|^2 + |t|^4 < 1\}.$$

If we use the group law to translate this ball so that it has center at $c + i0$, $c \in \mathbb{R}$, then we obtain

$$B_c = \{(z,t) : |z - c|^2 + |(t - \mathrm{Im}\,(cz))|^4 < 1\}.$$

Because of the condition the inequality imposes on the t variable, we find that B_c is rotated.

For the purposes of harmonic analysis, we may use an axiomatic structure that is in the spirit of some ideas of K. T. Smith [**SMI**]. Smith's ideas were developed further in [**HOR5**], [**COW1**], [**COW2**], and [**CHR1**]. Those sources have had an influence on the way that we have recorded the four axioms below. The resulting geometry serves as a substitute for the simple geometric structure of Euclidean boxes.

We note in passing that the present discussion is close in spirit to the notion of "space of homogeneous type" that is treated in more detail in Sections 2.3, 2.6. The theory of spaces of homogeneous type proved to be quite prescient, for when it was invented there were few interesting examples of these spaces available

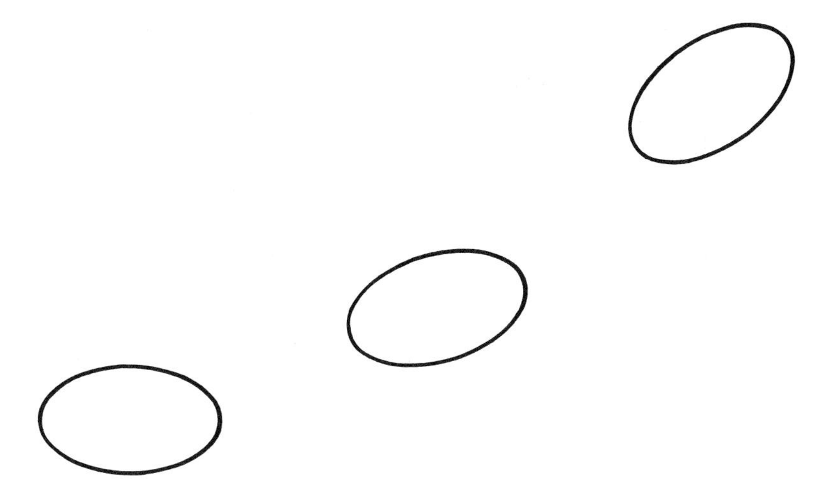

FIGURE 1.10

(the best were examples of Lie groups, but these certainly have translations and a maximal torus, therefore are in many ways too much like Euclidean space). Harmonic analysis on the boundary of a domain in $\mathbb{C}^n$ is an ideal venue in which to use the ideas of homogeneous type. This latter assertion will be developed in ensuing sections.

Let X be a topological space that is equipped with a measure μ and a family of balls $B(x,r)$ (for each $x \in X$ and each $r > 0$) that satisfies

a: $0 < \mu(B(x,r)) < \infty$ for each $x \in X$ and $r > 0$;

a': If $x \in X$ and $0 < r \leq s$ then $B(x,r) \subseteq B(x,s)$;

a": If $x \in X$ and $r > 0$ then $x \in B(x,r)$;

b: There is a number $C_1 > 1$ such that for any $x \in X, r > 0$, $\mu(B(x,2r)) \leq C_1\mu(B(x,r))$;

c: There is a number $C_2 > 1$ such that for any $x, y \in X, r \geq s > 0$, if $B(x,r) \cap B(y,s) \neq \emptyset$ then $B(x,C_2r) \supseteq B(y,s)$.

In many situations Axioms **a'** and **a"** can be omitted, and Axiom **c** can serve as a substitute.

On the Heisenberg group $\mathbb{H}_1$, one verifies properties **a** and **b** by checking that $\mu(B(x,r)) \approx r^4$. Properties **a'** and **a"** are obvious. Property **c** is a bit trickier. One checks that, although the major axis of the ball $B(y,s)$ is rotated with respect to the major axis of $B(y,s)$, the angle of rotation is $\mathrm{Tan}^{-1} r$. This, and some trigonometry, suffices to show that C_2 exists.

These axioms are sufficient to prove a covering theorem of the following sort:

THEOREM 1.2.1. *Let K be a compact subset of X. Assume that the balls $B(x_\alpha, r_\alpha)$ cover K. Then there is a subcollection*

$$B(x_{\alpha_1}, r_{\alpha_1}), \ldots, B(x_{\alpha_k}, r_{\alpha_k})$$

of pairwise disjoint balls such that $\{B(x_{\alpha_j}, C_2 r_{\alpha_j})\}_{j=1}^k$ *covers* K.

PROOF. Since K is compact, we may as well assume that we are beginning with a finite covering. Let $B(x_{\alpha_1}, r_{\alpha_1})$ be the largest of the balls (which is not necessarily unique). Next let $B(x_{\alpha_2}, r_{\alpha_2})$ be the largest of the remaining balls (not necessarily unique) which is disjoint from the first ball. At the kth step one chooses the largest ball that is disjoint from the first $k-1$ balls. Since there are only finitely many balls, the process eventually stops.

Thus we have obtained a pairwise disjoint subcollection. To check the other conclusion, let $B(x_\alpha, r_\alpha)$ be one of the balls from the original (finite) collection. If this is one of the balls that was selected, then of course it is covered by $\{B(x_{\alpha_j}, C_2 r_{\alpha_j})\}$. If it is not one of the selected balls, then there is a *first* selected ball $B(x_{\alpha_j}, r_{\alpha_j})$ that intersects $B(x_\alpha, r_\alpha)$ (otherwise we would have selected $B(x_\alpha, r_\alpha)$). By the way that we selected the balls, it must be that $r_{\alpha_j} \geq r_\alpha$. But then, by axiom **c**, it holds that $B(x_{\alpha_j}, C_2 r_{\alpha_j}) \supseteq B(x_\alpha, r_\alpha)$. Thus the balls $B(x_{\alpha_j}, C_2 r_{\alpha_j})$ cover all the original balls, and hence K itself. That completes the proof. ■

As will be explained in Chapter 2, the basic covering lemma just proved—together with axioms **a** and **b**—allows one to verify the boundedness of the Hardy-Littlewood operator

$$Mf(x) = \sup_{r>0} \frac{1}{\mu(B(x,r))} \int_{B(x,r)} |f(t)| d\mu(t)$$

on $L^p(X, d\mu)$, $1 < p < \infty$. One may prove an analogue of the Whitney decomposition and an analogue of the Calderón-Zygmund theorem for singular integrals. See [**COW1**].

The Heisenberg group is a prototype for the natural geometry of the boundary of a strongly pseudoconvex domain. The chief fact about a strongly pseudoconvex boundary is that the first commutator of the (real) basis elements for the complex tangent directions has a non-zero complex normal component. More precisely, let $\Omega = \{z \in \mathbb{C}^2 : \rho(z) < 0\}$ be a domain in complex dimension two. Define

$$L = \frac{\partial \rho}{\partial z_1} \frac{\partial}{\partial z_2} - \frac{\partial \rho}{\partial z_2} \frac{\partial}{\partial z_1}.$$

One checks directly that $L\rho \equiv 0$. Let $\bar{L}$ be the conjugate of L. By a classical formula of Cartan (see [**KRA1**]), one sees that $[L, \bar{L}]$ is just the determinant of the Levi form times T (the missing real tangential direction—or the complex normal direction) plus certain error terms consisting of multiples of L and $\bar{L}$. This is analogous to the commutator relations in the Heisenberg group. It follows (after some calculations) that the natural geometry of a strongly pseudoconvex boundary involves nonisotropic balls that have dimension $\sqrt{r}$ in complex tangential directions and dimension r in the complex normal direction. This geometry will be made even clearer when we treat the "finite type" situation below. [Another way to see this last assertion is by modeling the balls on sublevel sets of either the Szegö kernel or the Bergman kernel. According to the asymptotic expansions of Fefferman [**FEF**] and Boutet de Monvel/Sjöstrand [**BMS**] (see also [**EPM**], [**FAK**], [**HAG**]), the Bergman (resp. the Szegö) kernel $K(z,\zeta)$ on a strongly pseudoconvex domain, in suitable local coordinates near the boundary, has an asymptotic expansion of the form

$$K(z,\zeta) = K_B(z,\zeta) + \text{ lower order terms,}$$

where K_B is the Bergman kernel (resp. the Szegö kernel) for the ball. The singularity of the Bergman (resp. the Szegö) kernel on the ball is a power of $(1 - z\cdot\bar{\zeta})$, and sets of the form $\{\zeta : |1 - z\cdot\bar{\zeta}| < A\}$ have the parabolic geometry that we have been discussing.]

Now let us turn to the main topic of this section and the next: domains of finite type. It requires a bit of machinery to describe these domains.

Let us begin with the simplest domain in $\mathbb{C}^n$—the ball. Let $P \in \partial B$. No complex line can have geometric order of contact with ∂B at P exceeding 2. That is, a complex line may pass through P and also be tangent to ∂B at P, but it can do no better. The differential geometric structures disagree at the level of second derivatives. The number 2 is called the "order of contact" of the complex line with ∂B.

Strong pseudoconvexity can be viewed as the correct biholomorphically invariant version of the phenomenon described in the last paragraph: no analytic disc can osculate to better than first order tangency at a strongly pseudoconvex boundary point. In fact the positive definiteness of the Levi form provides the obstruction that makes this statement true. Let us sketch a proof:

Suppose that $P \in \partial\Omega$ is a point of strong pseudoconvexity and that ρ is a defining function whose complex Hessian at P is positive definite on all of $\mathbb{C}^n$. We may normalize coordinates so that $P = 0$ and $(1, 0, \ldots, 0)$ is the unit outward normal at P. We write

$$\begin{aligned}
\rho(w) &= \rho(P) + \sum_{j=1}^n \frac{\partial\rho}{\partial z_j}(P)w_j + \frac{1}{2}\sum_{j,k=1}^n \frac{\partial^2\rho}{\partial z_j\partial z_k}(P)w_jw_k \\
&\quad + \sum_{j=1}^n \frac{\partial\rho}{\partial \bar{z}_j}(P)\bar{w}_j + \frac{1}{2}\sum_{j,k=1}^n \frac{\partial^2\rho}{\partial \bar{z}_j\partial \bar{z}_k}(P)\bar{w}_j\bar{w}_k \\
&\quad + \sum_{j,k=1}^n \frac{\partial^2\rho}{\partial z_j\partial \bar{z}_k}(P)w_j\bar{w}_k + o(|w|^2) \\
&= 2\mathrm{Re}\left\{\sum_{j=1}^n \frac{\partial\rho}{\partial z_j}(P)w_j + \frac{1}{2}\sum_{j,k=1}^n \frac{\partial^2\rho}{\partial z_j\partial z_k}(P)w_jw_k\right\} \\
&\quad + \sum_{j,k=1}^n \frac{\partial^2\rho}{\partial z_j\partial \bar{z}_k}(P)w_j\bar{w}_k + o(|w|^2)
\end{aligned}$$

$$= 2\mathrm{Re}\left\{ w_1 + \frac{1}{2}\sum_{j,k=1}^{n} \frac{\partial^2 \rho}{\partial z_j \partial z_k}(P) w_j w_k \right\}$$
$$+ \sum_{j,k=1}^{n} \frac{\partial^2 \rho}{\partial z_j \partial \bar{z}_k}(P) w_j \bar{w}_k + o(|w|^2) \qquad (1)$$

Define the mapping $w = (w_1, \ldots, w_n) \mapsto w' = (w_1', \ldots, w_n')$ by

$$\begin{aligned}
w_1' &= \Phi_1(w) &&= w_1 + \frac{1}{2}\sum_{j,k=1}^{n} \frac{\partial^2 \rho}{\partial z_j \partial z_k}(P) w_j w_k \\
w_2' &= \Phi_2(w) &&= w_2 \\
&\quad\cdot &&\quad\cdot \\
&\quad\cdot &&\quad\cdot \\
&\quad\cdot &&\quad\cdot \\
w_n' &= \Phi_n(w) &&= w_n.
\end{aligned}$$

By the implicit function theorem, we see that for w sufficiently small this is a well-defined invertible holomorphic mapping on a small neighborhood W of $P = 0$. Then equation (1) tells us that, in the coordinate w', the defining function becomes

$$\hat{\rho}(w') = 2\mathrm{Re}\, w_1' + \sum_{j,k=1}^{n} \frac{\partial^2 \rho}{\partial z_j' \partial \bar{z}_k'}(P) w_j' \bar{w}_k' + o(|w'|^2).$$

Thus the *real* Hessian at P of the defining function $\hat{\rho}$ is precisely the Levi form; and the latter is positive definite by our hypothesis. In other words, the only second order terms in the Taylor expansion of ρ about P are the mixed terms occuring in the complex Hessian (that is, there are no $\partial^2/\partial z_j \partial z_k$ terms and no $\partial^2/\partial \bar{z}_j \partial \bar{z}_k$ terms).

Let $\phi : D \to \mathbb{C}^n$ be an analytic disc that is tangent to $\partial\Omega$ at $P : \phi(0) = P$ and $\phi'(0) \neq 0$. The tangency means that

$$\mathrm{Re}\left(\sum_{j=1}^{n} \frac{\partial \rho}{\partial z_j}(P) \phi_j'(0) \right) = 0.$$

If we expand $\hat{\rho} \circ \phi(\zeta)$ in a Taylor expansion about $\zeta = 0$ then the zero and first order terms vanish. As a result, for ζ small,

$$\rho \circ \phi(\zeta) = \left[\sum_{j,k=1}^{n} \frac{\partial^2 \rho}{\partial z_j \partial \bar{z}_k}(P) \phi_j'(0) \bar{\phi}_k'(0) \right] |\zeta|^2 + o(|\zeta|^2).$$

This is

$$\geq C \cdot |\zeta|^2$$

for ζ small. Thus we have an explicit lower bound, in terms of the eigenvalues of the Levi form, for the order of contact of the image of ϕ with $\partial\Omega$.

EXAMPLE 1.2.2. Let $0 \le m \in \mathbb{Z}$. Set

$$\Omega = \Omega_m = \{(z_1, z_2) \in \mathbb{C}^2 : \rho(z_1, z_2) = -1 + |z_1|^2 + |z_2|^{2m} < 0\}.$$

Let $P = (1,0) \in \partial\Omega$. Let $\phi : D \to \mathbb{C}^2$ be a non-singular analytic disc that is tangent to $\partial\Omega$ at P : that is, $\phi(0) = P, \phi'(0) \neq 0, [\rho \circ \phi]'(0) = 0$. We may in fact assume, after a reparametrization, that $\phi'(0) = (0,1)$. Then

$$\phi(\zeta) = (1 + 0 \cdot \zeta + \mathcal{O}(\zeta^2), \zeta + \mathcal{O}(\zeta^2)). \tag{2}$$

What is the greatest order of contact that such a disc ϕ can have with $\partial\Omega$?

To answer this question, we let

$$\rho(z) = -1 + |z_1|^2 + |z_2|^{2m}$$

be the defining function for the domain.

Obviously the disc $\phi(\zeta) = (1, \zeta)$ has order of contact $2m$ at $P = (1,0)$, for

$$\rho \circ \phi(\zeta) = |\zeta|^{2m} = \mathcal{O}\big(|(1,\zeta) - (1,0)|^{2m}\big).$$

The question is whether we can improve upon this estimate with a different curve ϕ. Since all curves ϕ under consideration must have the form (2), we calculate that

$$\begin{aligned} \rho \circ \phi(\zeta) &= -1 + \left|1 + \mathcal{O}(\zeta^2)\right|^2 + \left|\zeta + \mathcal{O}(\zeta^2)\right|^{2m} \\ &= -1 + \left[\left|1 + \mathcal{O}(\zeta^2)\right|^2\right] + \left[|\zeta|^{2m}\left|1 + \mathcal{O}(\zeta)\right|^{2m}\right]. \end{aligned}$$

The second expression in brackets is essentially $|\zeta|^{2m}$ so if we wish to improve on the order of contact then the first term in [ ] must cancel it. But that first term would be $|1 + c\zeta^m + \cdots|^2$. The resulting term of order $2m$ would be positive and in fact would *not* cancel the second. We conclude that $2m$ is the best order of contact for holomorphic curves with $\partial\Omega$ at P. We will say that $P \in \partial\Omega_m$ is of "geometric type $2m$."

Now consider the domain Ω_m from the analytic viewpoint. Define

$$\begin{aligned} L &= \frac{\partial\rho}{\partial z_1}\frac{\partial}{\partial z_2} - \frac{\partial\rho}{\partial z_2}\frac{\partial}{\partial z_1} \\ &= \bar{z}_1 \frac{\partial}{\partial z_2} - m z_2^{m-1} \bar{z}_2^m \frac{\partial}{\partial z_1} \end{aligned}$$

and

$$\begin{aligned} \bar{L} &= \frac{\partial\rho}{\partial \bar{z}_1}\frac{\partial}{\partial \bar{z}_2} - \frac{\partial\rho}{\partial \bar{z}_2}\frac{\partial}{\partial \bar{z}_1} \\ &= z_1 \frac{\partial}{\partial \bar{z}_2} - m \bar{z}_2^{m-1} z_2^m \frac{\partial}{\partial \bar{z}_1}. \end{aligned}$$

One can see from their very definition, or can compute directly, that both of these vector fields are tangent to $\partial\Omega$. That is to say, $L\rho \equiv 0$ and $\bar{L}\rho \equiv 0$. Of course the commutator of two tangential vector fields must still be tangential *in the sense of real geometry*. That is, $[L, \bar{L}]$ must lie in the three dimensional real tangent space to $\partial\Omega$ at each point of $\partial\Omega$. However there is no guarantee that this commutator must lie in the *complex* tangent space. In general it will not. Consider the case $m = 1$—the ball. A calculation reveals that, at the point $P = (1, 0)$,

$$[L, \bar{L}] \equiv L\bar{L} - \bar{L}L = -i\frac{\partial}{\partial y_1}.$$

This vector is indeed tangent to the boundary of the ball at P (it annihilates the defining function); but it is a scalar multiple of $-J(\partial/\partial x_1)$, where J is the complex structure tensor (and $\partial/\partial x_1$ is the usual Euclidean normal vector). Therefore it is what we call *complex normal.*

The reason that, on the ball, it takes only a commutator of order one to escape the complex tangent and have a component in the complex normal direction is that the ball is strongly pseudoconvex. On our domain Ω_m, at the point $P = (1, 0)$, it requires a commutator

$$[L, [\bar{L}, \ldots [L, \bar{L}] \ldots]]$$

of length $2m - 1$ (that is, a total of $2m$ vector fields L's and $\bar{L}$'s) to have a component in the complex normal direction. We say that P is of "analytic type $2m$."

Thus, in this simple example, a point of geometric type $2m$ is of analytic type $2m$. ■

1.3. Finite Type. Now we shall give the formal definitions of geometric type and of analytic type for a point in the boundary of a smoothly bounded domain $\Omega \subseteq \mathbb{C}^2$. We fix the dimension to be two until further notice. The main result will be that the two notions are equivalent.

We begin by defining a gradation of vector fields which will be the basis for our definition of analytic type. Throughout this section $\Omega = \{z \in \mathbb{C}^2 : \rho(z) < 0\}$ and ρ is C^∞. If $P \in \partial\Omega$ then we may make a change of coordinates so that $\partial\rho/\partial z_2(P) \neq 0$. Define the holomorphic vector field

$$L = \frac{\partial\rho}{\partial z_1}\frac{\partial}{\partial z_2} - \frac{\partial\rho}{\partial z_2}\frac{\partial}{\partial z_1}$$

and the conjugate holomorphic vector field

$$\bar{L} = \frac{\partial\rho}{\partial \bar{z}_1}\frac{\partial}{\partial \bar{z}_2} - \frac{\partial\rho}{\partial \bar{z}_2}\frac{\partial}{\partial \bar{z}_1}.$$

Both L and $\bar{L}$ are tangent to the boundary because $L\rho = 0$ and $\bar{L}\rho = 0$. They are both non-vanishing near P by our normalization of coordinates.

The real and imaginary parts of L (equivalently of $\bar{L}$) generate (over the ground field $\mathbb{R}$) the complex tangent space to $\partial\Omega$ at all points near P. The vector field L alone generates the space of all holomorphic tangent vector fields and $\bar{L}$ alone generates the space of all conjugate holomorphic tangent vector fields.

DEFINITION 1.3.1. Let $\mathcal{L}_1$ denote the module, over the ring of C^∞ functions, generated by L and $\bar{L}$. Inductively, $\mathcal{L}_\mu$ denotes the module generated by $\mathcal{L}_{\mu-1}$ and all commutators of the form $[F, G]$ where $F \in \mathcal{L}_1$ and $G \in \mathcal{L}_{\mu-1}$.

Clearly $\mathcal{L}_1 \subseteq \mathcal{L}_2 \subseteq \cdots$. Each $\mathcal{L}_\mu$ is closed under conjugation. *It is not generally the case that $\cup_\mu \mathcal{L}_\mu$ is the entire three dimensional tangent space at each point of the boundary.* A counterexample is provided by

$$\Omega = \{z \in \mathbb{C}^2 : |z_1|^2 + 2e^{-1/|z_2|^2} < 1\}$$

and the point $P = (1, 0)$.

DEFINITION 1.3.2. Let $\Omega = \{\rho < 0\}$ be a smoothly bounded domain in $\mathbb{C}^2$ and let $P \in \partial\Omega$. We say that $\partial\Omega$ is of *finite analytic type $m \geq 2$ at P* if $\langle \partial\rho(P), F(P)\rangle = 0$ for all $F \in \mathcal{L}_{m-1}$ while $\langle \partial\rho(P), G(P)\rangle \neq 0$ for some $G \in \mathcal{L}_m$. In this circumstance we call P a point of analytic type m.

Notice that the condition $\langle \partial\rho(P), G(P)\rangle \neq 0$ is just an elegant way of saying that the vector $G(P)$ has non-zero component in the complex normal direction. This is plainly an open condition. Any point of the form $(e^{i\theta}, 0)$ in the boundary of $\{(z_1, z_2) : |z_1|^2 + |z_2|^{2m} < 1\}$ is of finite analytic type $2m$. Any point of the form $(e^{i\theta}, 0)$ in the boundary of $\Omega = \{z \in \mathbb{C}^2 : |z_1|^2 + 2e^{-1/|z_2|^2} < 1\}$ is *not* of finite analytic type. We say that such a point is of *infinite analytic type.*

There is ample opportunity for confusion when one is first learning to treat points of finite analytic type in the boundary of a domain in $\mathbb{C}^2$. Naively, it sounds as though certain tangent vector fields commute with each other finitely many times and suddenly become normal. This is not the case. First, all vector fields under consideration are tangential in the real variable sense. The issue is whether the vector fields have a component in the complex normal direction, that is the direction that is i times the usual Euclidean normal direction.

It is enlightening to consider a specific situation in coordinates. Fix a point $P \in \partial\Omega$ and assume that coordinates have been selected so that $P = 0$ and so that the stand real Euclidean normal direction is the $\operatorname{Re} z_1 = x_1$ direction. Then the tangent vector field L will have the form

$$L = \alpha(z)\frac{\partial}{\partial z_1} + \beta(z)\frac{\partial}{\partial z_2}.$$

Since L is complex tangential at $P = 0$, it holds that $\alpha(0) = 0$. Now each time we calculate a commutator $[L, \bar{L}], [L, [L, \bar{L}]]$, etc., the coefficient α will be differentiated. If α only vanishes to finite order at 0, then enough commutators of L and $\bar{L}$ will cause the coefficient of $\partial/\partial z_1$ in that commutator to be non-vanishing (once a non-vanishing derivative of α is reached). Thus this commutator will have a non-zero component in the complex normal direction. It will follow that $\partial\Omega$ is of finite type at P.

We encourage the reader to work out the ideas just described in the case of the ellipsoid E_m described in Example 1.2.2 at the boundary point $(1, 0)$. Then it can be seen, with each successive commutator, how the order of vanishing of the coefficient $\partial/\partial z_1$ drops with each successive commutator.

Now we turn to a precise definition of finite geometric type. If P is a point in the boundary of a smoothly bounded domain then we say that an analytic disc $\phi : D \to \mathbb{C}^2$ is a *non-singular disc tangent to* $\partial\Omega$ *at* P if $\phi(0) = P, \phi'(0) \neq 0$, and $(\rho \circ \phi)'(0) = 0$.

DEFINITION 1.3.3. Let $\Omega = \{\rho < 0\}$ be a smoothly bounded domain and $P \in \partial\Omega$. Let m be a positive integer. We say that $\partial\Omega$ is of *finite geometric type* m at P if the following condition holds: there is a non-singular disc $\phi : D \to \mathbb{C}^n, \phi(0) = P, \phi'(0) \neq 0$, such that, for small ζ,

$$|\rho \circ \phi(\zeta)| \leq C|\zeta|^m;$$

however there is no non-singular disc ψ tangent to $\partial\Omega$ at P such that, for small ζ,

$$|\rho \circ \psi(\zeta)| \leq C|\zeta|^{(m+1)}.$$

In this circumstance we call P a point of *finite geometric type* m.

We invite the reader to reformulate the definition of geometric finite type in terms of the order of vanishing of ρ restricted to the image of ϕ.

In what follows, we shall sometimes refer to this definition of geometric type as "regular type", since it is formulated in terms of non-singular analytic discs. In higher dimensions it is appropriate to allow *singular discs*, for reasons that shall be explained below.

Our first main result is the following theorem (see [**KON1**], [**BLG**]):

THEOREM 1.3.4 (KOHN, BLOOM/GRAHAM). *Let* $\Omega = \{\rho < 0\} \subseteq \mathbb{C}^2$ *be smoothly bounded and* $P \in \partial\Omega$. *Then* P *is of finite geometric type* $m \geq 2$ *if and only if it is of finite analytic type* m.

PROOF. We may assume that $P = 0$. Write ρ in the form

$$\rho(z) = 2\text{Re}\, z_2 + f(z_1) + \mathcal{O}(|z_1 z_2| + |z_2|^2).$$

Notice that

$$L = \frac{\partial f}{\partial z_1}\frac{\partial}{\partial z_2} - \frac{\partial}{\partial z_1} + (\text{error terms}).$$

Here the error terms arise from differentiating $\mathcal{O}(|z_1 z_2| + |z_2|^2)$. Now the best order of contact of a one dimensional non-singular complex variety with $\partial\Omega$ at 0 equals the order of contact of the variety $\zeta \mapsto (\zeta, 0)$ with $\partial\Omega$ at 0 which is just the order of vanishing of f at 0.

On the other hand,

$$\begin{aligned}[L,\bar{L}] &= \left[-\frac{\partial^2 \bar{f}}{\partial z_1 \partial \bar{z}_1}\frac{\partial}{\partial \bar{z}_2}\right] - \left[-\frac{\partial^2 f}{\partial \bar{z}_1 \partial z_1}\frac{\partial}{\partial z_2}\right] \\ &\quad +(\text{error terms}) \\ &= 2i\,\mathrm{Im}\left[\frac{\partial^2 f}{\partial \bar{z}_1 \partial z_1}\frac{\partial}{\partial z_2}\right] \\ &\quad +(\text{error terms}).\end{aligned}$$

Inductively, we see that a commutator of m vector fields chosen from $L, \bar{L}$ will consist of (real or imaginary parts of) m^{th} order derivatives of f times $\partial/\partial z_2$ plus the usual error terms. And the pairing of such a commutator with $\partial\rho$ at 0 is just the pairing of that commutator with dz_2; in other words it is just the coefficient of $\partial/\partial z_2$. We see that this number is non-vanishing as soon as the corresponding derivative of f is non-vanishing. Thus the analytic type of 0 is just the order of vanishing of f at 0.

Since both notions of type correspond to the order of vanishing of f, we are done. ∎

From now on, when we say "finite type" (in dimension two), we can mean either the geometric or the analytic definition.

We say that a domain $\Omega \subseteq \mathbb{C}^2$ is of *finite type* if there is a number M such that every boundary point is of finite type not exceeding M.

Some general remarks about finite type are in order. From the very definition, we see that the set of finite type points in a smooth boundary form an open set. In fact, for any m, the set of points of type $\leq m$ form an open set (one of those rare instances in mathematics where $\leq$ gives an open set and $>$ gives a closed set). It follows (exercise) that if Ω is smoothly bounded and if each point of $\partial\Omega$ has finite type then there exists a positive integer M so that every boundary point has type not exceeding M. Thus such an Ω is of finite type as defined above.

What is slightly less obvious is that it is impossible for a relatively open set $U \subseteq \partial\Omega \subseteq \mathbb{C}^2$ to have the property that $2 < \text{type}(P) \leq m < \infty$ for all $P \in U \cap \partial\Omega$. To see this, one observes that this hypothesis would imply that the Levi form has rank 0 at each point of $U \cap \partial\Omega$ (since each point is *not* strongly pseudoconvex) hence, by a standard result, that $U \cap \partial\Omega$ is foliated by complex analytic varieties (see [**KRA1**]). But then each point has type ∞.

An analogue of this last assertion is true in dimensions three and higher, but is harder to see. Catlin's theory of multi-type [**CAT2**] implies that, in a pseudoconvex, finite type boundary, the weakly pseudoconvex points lie in a stratified collection of submanifolds. Thus they form a nowhere dense set of zero measure. We will say more about finite type in dimensions three and greater in what follows.

The most important, but by no means the only, examples of domains of finite type are the bounded domains with real analytic boundary. It is a non-trivial matter to verify this assertion. The first proof appears in [**DIF2**]. A fresh approach to these ideas appears in [**DAN1**].

Analysis on finite type domains in $\mathbb{C}^2$ has recently become a matter of great interest. Analysis on a finite type domain in complex dimension two is tractable because a finite type point has a "product structure" geometry (the boundary balls associated to such a point of type m have dimensions r by r^m). What is new, and different from the strongly pseudoconvex case, is that the geometry varies from point to point. This variation is upper semi-continuous (exercise), but it requires a upper semi-continuity of type rather complex weighted sum of powers of r with coefficients given by the commutators of L and $\bar{L}$ to truly control the geometry. We now briefly recall the relevant ideas from [**NRSW**]. Similar ideas appear in [**CAT4**] and [**MCN1**].

Let ρ be a defining function for the finite type domain Ω, normalized so that $\nabla\rho \equiv 1$ on $\partial\Omega$. As above, we set

$$L = \frac{\partial\rho}{\partial z_1}\frac{\partial}{\partial z_2} - \frac{\partial\rho}{\partial z_2}\frac{\partial}{\partial z_1}.$$

For a domain in complex dimension 2, this is the unique holomorphic tangent vector field. Also set

$$S = 2\left[\frac{\partial\rho}{\partial\bar{z}_1}\frac{\partial}{\partial z_1} + \frac{\partial\rho}{\partial\bar{z}_2}\frac{\partial}{\partial z_2}\right].$$

We let $T = \operatorname{Im} S$. Let $X_1, -X_2$ (the minus sign is introduced for reasons of signature that come up in applications) denote the real and imaginary parts of L. Then X_1, X_2, T are tangential vector fields that span the tangent space to $\partial\Omega$ at each point.

Now for each multi-index $(i_1, \ldots, i_k)$ with the i_j assuming only the values $1, 2$ we define functions $\lambda_{i_1,\ldots,i_k}, \alpha^1_{i_1,\ldots,i_k}, \alpha^2_{i_1,\ldots,i_k}$ using the equation

$$\Big[X_{i_k}, \big[\ldots[X_{i_2}, X_{i_1}]\ldots\big]\Big] = \lambda_{i_1,\ldots,i_k}T + \alpha^1_{i_1,\ldots,i_k}X_1 + \alpha^2_{i_1,\ldots,i_k}X_2.$$

For each index $\ell \geq 2$ we define a function

$$\Lambda_\ell(p) = \left[\sum |\lambda_{i_1,\ldots,i_k}(p)|^2\right]^{1/2}.$$

Here the sum is taken over all k-tuples with $2 \leq k \leq \ell$.

In our new notation, a point $p \in \partial\Omega$ is of type m if $\Lambda_2(p) = \cdots = \Lambda_{m-1}(p) = 0$ but $\Lambda_m(p) \neq 0$. A domain is said to be of *type m* if every point of the boundary is of type at most m. If Ω is a domain of type m, $p \in \partial\Omega$, and $\delta > 0$, then we define

$$\Lambda(p, \delta) = \sum_{j=2}^{m} \Lambda_j(p)\delta^j.$$

The quantity $\Lambda(p, \delta)$ is the device that Nagel, Rosay, Stein, and Wainger created to regulate the geometry of the boundary of a finite type domain. In case the domain is strongly pseudoconvex, then $\Lambda_2(p) \neq 0$ for every boundary point p. By continuity, the quantity is bounded from 0. As a result, when δ is small then the first term in the sum defining $\Lambda(p, \delta)$ swamps all the others. Therefore, in the strongly pseudoconvex case,

$$\Lambda(p, \delta) \approx \delta^2.$$

This is a concrete realization of the fact that the geometry on a strongly pseudoconvex hypersurface is uniformly parabolic.

We note that McNeal [**MCN1**] has a related quantity τ that is, in effect, the inverse of Λ. In many analytic contexts, it is more natural to use τ.

As previously noted, on a more general finite type hypersurface the geometry is considerably more complicated. Even in the simple example

$$\Omega = \{(z_1, z_2) : |z_1|^2 + |z_2|^4 < 1\}$$

the geometry degenerates from parabolic to quartic as a boundary point (z_1, z_2) passes from z_2 small and non-zero to $z_2 = 0$. The quantity $\Lambda(p, \delta)$ (obversely McNeal's parameter τ) measures this change.

Finite Type in Higher Dimensions

One natural generalization of the notion of geometric finite type from dimension two to dimensions three and higher is to consider orders of contact of $(n-1)$ dimensional complex manifolds with the boundary of a domain Ω at a point P. The definition of analytic finite type generalizes to higher dimensions almost directly (one deals with tangent vector fields $L_1, \ldots, L_{n-1}$ and $\bar{L}_1, \ldots, \bar{L}_{n-1}$ instead of just L and $\bar{L}$). It is a theorem of [**BLG**] that, with these definitions, geometric finite type and analytic finite type are the same in all dimensions.

This is an elegant result, and is entirely suited to certain questions of extension of CR functions and reflections of holomorphic mappings. However, it is not the correct indicator of when the $\bar{\partial}$- Neumann problem exhibits a gain in the Sobolev topology—that is, when the unique solution to the $\bar{\partial}$-Neumann problem is smoother in the Sobolev topology than is the data. In the late 1970's and early 1980's, John D'Angelo realized that a correct understanding of finite type in all dimensions requires sophisticated ideas from algebraic geometry—particularly the intersection theory of analytic varieties. And he saw that *non-singular varieties cannot tell the whole story.* An important sequence of papers, beginning with [**DAN2**], laid down the theory of domains of finite type in all dimensions. The complete story of this work, together with its broader mathematical context, appears in [**DAN1**]. David Catlin [**CAT1**], [**CAT2**], [**CAT3**] validated the significance of D'Angelo's work by proving that the $\bar{\partial}$- Neumann problem has a gain in the Sobolev topology near a point $P \in \partial\Omega$ if and only if the point P is of finite type in the sense of D'Angelo. [It is interesting to note that there are partial differential operators that exhibit a gain in the Sobolev topology but not in the Lipschitz topology—see Guan [**GUA**].]

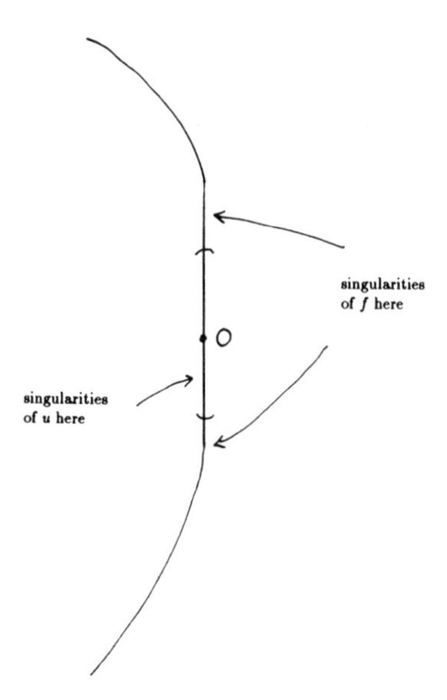

FIGURE 1.11

The point is that analytic structure in the boundary of a domain is an obstruction to regularity for the $\bar{\partial}$ problem. In fact if the boundary contains an analytic disc then it is possible for the equation $\bar{\partial}u = f$ to have data f that is C^∞ but so that there is no smooth solution u. Let us flesh out this assertion.

Consider a domain $\Omega \subseteq \mathbb{C}^2$ with a piece of boundary of the form $\operatorname{Re} z_2 = 0$ oriented such that $\operatorname{Re} z_2 < 0$ corresponds to the interior of the domain. Assume that the origin lies in the piece of boundary and let $U = \{z \in \mathbb{C}^2 : |z| < 4\delta\}$ be a neighborhood of the origin such that $U \cap \partial\Omega = U \cap \{z : \operatorname{Re} z_2 = 0\}$. The parameter δ is fixed once and for all. Let $\phi \in C_c^\infty(U)$ satisfy $\phi(z) \equiv 1$ when $|z| < \delta$, $\phi(z) \equiv 0$ when $|z| > 2\delta$. Set $f = \bar{\partial}[\phi/z_2] = (\bar{\partial}\phi)/z_2$.

Then f is a $\bar{\partial}$-closed $(0,1)$ form on Ω *that is smooth up to the portion of the boundary* where $\operatorname{Re} z_2 = 0$, $|z_1| < \delta$ and where $\operatorname{Re} z_2 = 0, |z_1| > 2\delta$. Nonetheless, the singularities of f, which arise for $\delta \le |z_1| \le 2\delta$ (see Figure 1.11), propagate outside the support of f for solutions of the equation $\bar{\partial}u = f$.

To see why this is true, let us formulate an exact statement. Suppose that $\bar{\partial}u = f$ on Ω and suppose that u is bounded up to that portion of the boundary given by

$$\{\operatorname{Re} z_2 = 0\} \cap \big[\{|z_1| < \delta\} \cup \{2\delta < |z_1| < 4\delta\}\big].$$

We derive a contradiction as follows: Define the analytic function $h \equiv u - \phi/z_2$ on Ω. For $\epsilon > 0$ and $|z_1| < \delta/2$ let us represent h by its Cauchy integral

$$h(z_1, \epsilon) = \frac{1}{2\pi i} \oint_{|\zeta|=3\delta} \frac{h(\zeta, \epsilon)}{\zeta - z_1}\, d\zeta. \tag{1}$$

With $\delta > 0$ fixed, let us examine this formula as $\epsilon \to 0^+$. By our hypothesis on u, it is bounded on the contour of integration. Also ϕ/z_2 vanishes on the contour of integration. And of course $|\zeta - z_1$ is bounded from zero on that contour. So the right side of (1) is bounded. For the specified range of z_1, the function $u(z_1, \epsilon)$ is also bounded as $\epsilon \to 0^+$. However the function ϕ/z_2 blows up at the points (z_1, ϵ) as $\epsilon \to 0^+$. That is a contradiction.

We see that the presence of a variety in the boundary can lead to lack of boundary regularity for the $\bar{\partial}$ problem. A comment is in order. There is no logical reason to distinguish the part of the boundary that lies *inside* the annular region on which f is singular from that part which lies *outside.* Therefore we have formulated our propagation of singularities result as a dialectic. However the same set of ideas can be used to prove several other "propagation of singularities" statements. The example first appeared in [**KON1**].

In higher dimensions, matters become technical rather quickly. Therefore we shall content ourselves with a primarily descriptive treatment of this material. The interesting new feature in dimensions $n \geq 3$ is that it requires algebro-geometric ideas, particularly intersection theory and ideal theory, to make sharp statements about type. Thus, taking into account what we know in all dimensions, there are three (or four) versions of type: contact type using non-singular varieties, contact type using singular varieties, commutator type, and ideal (or D'Angelo) type. As of this writing, there is no "analytic" description in higher dimensions of finite type using commutators of vector fields. [*However*: heartening progress in this matter has been made in the case of convex domains. See [**MCN1**], [**MCN2**], [**MCN3**], [**BOS**], [**YU2**].] Jiye Yu has also done some recent work that begins to identify those domains for which all the notions of type coincide. Other results along these lines are due to McNeal. We know that the notion of finite type that we are about to describe is the right one for the study of the $\bar{\partial}$-Neumann problem because (i) it enjoys certain important semi-continuity properties (to be discussed below) that other notions of finite type do not, and (ii) Catlin's theorem shows that the definition meshes perfectly with fundamental concepts like the regularity of the $\bar{\partial}$- Neumann problem.

We would be remiss not to at least mention J. J. Kohn's theory of type, in all dimensions, that depends on ideals of forms. It is believed that this theory is essentially equivalent to D'Angelo's theory, but the details remain to be worked out. We refer the reader to [**KON3**] and [**DAN4**], [**DAN5**], [**DAN6**].

However many fundamental questions of hard analysis on finite type domains in dimensions $n \geq 3$ remain open. In particular, the natural geometry of a finite point in $\mathbb{C}^3$ (for instance) is not understood—it is certainly much more complicated than a product. Also the semi-continuity in this situation is much more subtle.

What we learn next is that the order of contact of analytic varieties stratifies the insight provided by the propagation of singularities example into degrees, so that one may make fairly precise statements about the "gain" of the $\bar{\partial}$ problem in terms of the order of contact of varieties at the boundary.

Let us begin by introducing some notation. Let $U \subseteq \mathbb{C}^n$ be an open set. A subset $V \subseteq U$ is called a *variety* if for each $P \in V$ there is a neighborhood $\mathcal{W}_P$ such that there are holomorphic functions $f_1, \ldots, f_k$ on U and $\mathcal{W}_P \cap V = \{z \in \mathcal{W}_P : f_1(z) = \cdots = f_k(z) = 0\}$. A variety is called *irreducible* if it cannot be written as the union of proper non-trivial subvarieties. One dimensional varieties are particularly easy to work with because they can be parametrized (see [**GUN**]):

PROPOSITION 1.3.5. *Let $V \subseteq \mathbb{C}^n$ be an irreducible one dimensional complex analytic variety. Let $P \in V$. There is a neighborhood W of P and a holomorphic mapping $\phi : D \to \mathbb{C}^n$ such that $\phi(0) = P$ and the image of ϕ is $W \cap V$. When this parametrization is in place then we refer to the variety as a holomorphic curve.*

In general, we cannot hope that the parametrization ϕ will satisfy (nor can it be arranged that) $\phi'(0) \neq 0$. As a simple example, consider the variety

$$V = \{z \in \mathbb{C}^2 : z_1^2 - z_2^3 = 0\}.$$

Then the most natural parametrization for V is $\phi(\zeta) = (\zeta^3, \zeta^2)$. Notice that $\phi'(0) = 0$ and there is no way to reparametrize the curve to make the derivative non-vanishing. This is because the variety has a singularity—a cusp—at the point $P = 0$.

DEFINITION 1.3.6. Let f be a scalar-valued holomorphic function of a complex variable and P a point of its domain. The *multiplicity* of f at P is defined to be the least positive integer k such that the kth derivative of f does not vanish at P. If m is that multiplicity then we write $v_P(f) = v(f) = m$.

If ϕ is instead a vector-valued holomorphic function of a complex variable then its multiplicity at P is defined to be the minimum of the multiplicities of its entries. If that minimum is m then we write $v_P(\phi) = v(\phi) = m$.

For the remainder of this section we will exclusively calculate the multiplicities of holomorphic curves $\phi(\zeta)$ at $\zeta = 0$.

For example, the function $\zeta \mapsto \zeta^2$ has multiplicity 2 at 0; the function $\zeta \mapsto \zeta^3$ has multiplicity 3 at 0. Therefore the curve $\zeta \mapsto (\zeta^2, \zeta^3)$ has multiplicity 2 at 0.

If ρ is the defining function for a domain Ω then of course the boundary of Ω is given by the equation $\rho = 0$. The idea is to consider the *pullback* of the function ρ under a curve ϕ :

DEFINITION 1.3.7. Let $\phi : D \to \mathbb{C}^n$ be a holomorphic curve and ρ the defining function for a hypersurface $M \subseteq \mathbb{C}^n$ (usually, but not necessarily, the boundary of a domain). Then the pullback of ρ under ϕ is the function $\phi^*\rho(\zeta) = \rho \circ \phi(\zeta)$.

DEFINITION 1.3.8. Let M be a real hypersurface and $P \in M$. Let ρ be a defining function for M in a neighborhood of P. We say that P is a point of *finite type* (or finite 1-type) if there is a constant $C > 0$ such that

$$\frac{v(\phi^*\rho)}{v(\phi)} \leq C$$

whenever ϕ is a non-constant one-dimensional holomorphic curve through P such that $\phi(0) = P$.

The infimum of all such constants C is called the *type* (or 1-type) of P. It is denoted by $\Delta(M, P) = \Delta_1(M, P)$.

This definition is algebro-geometric in nature. We now offer a more geometric condition that is equivalent to it.

PROPOSITION 1.3.9. *Let P be a point of the hypersurface M. Let $\mathcal{E}_P$ be the collection of one-dimensional complex varieties passing through P. Then we have*

$$\Delta(M, P) = \sup_{V \in \mathcal{E}_P} \sup_{a>0} \left\{ a \in \mathbb{R}^+ : \lim_{V \ni z \to P} \frac{dist(z, M)}{|z - P|^a} \text{ exists and is finite} \right\}.$$

We leave the proof of the Proposition as an exercise. Notice that its statement is attractive in that it gives a characterization of finite type that makes no reference to a defining function. The proposition, together with the material in the first part of this section, motivates the following definition:

DEFINITION 1.3.10. Let P be a point of the hypersurface M. Let $\mathcal{R}_P$ be the collection of non-singular one-dimensional complex varieties passing through P (that is, we consider curves $\phi : D \to \mathbb{C}^n$, $\phi(0) = P$, $\phi'(0) \neq 0$). Then we define

$$\begin{aligned} \Delta^{reg}(M, P) &= \Delta_1^{reg}(M, P) \\ &= \sup_{V \in \mathcal{R}_P} \sup_{a>0} \left\{ a \in \mathbb{R}^+ : \lim_{V \ni z \to P} \frac{\operatorname{dist}(z, M)}{|z - P|^a} \text{ exists} \right\}. \end{aligned}$$

The number $\Delta_1^{reg}(M, P)$ measures order of contact of non-singular complex curves (i.e. one dimensional complex analytic *manifolds*) with M at P. By contrast, $\Delta_1(M, P)$ looks at all curves, both singular and non-singular. Obviously $\Delta_1^{reg}(M, P) \leq \Delta_1(M, P)$. The following example of D'Angelo shows that the two concepts are truly different:

EXAMPLE 1.3.11. Consider the hypersurface in $\mathbb{C}^3$ with defining function given by

$$\rho(z) = 2\text{Re}\, z_3 + |z_1^2 - z_2^3|^2.$$

Let the point P be the origin. Then we have

- We may calculate that $\Delta_1^{reg}(M, P) = 6$. We determine this by noticing that the z_3 direction is the normal direction to M at P hence any tangent curve must have the form

$$\zeta \mapsto (a(\zeta), b(\zeta), \mathcal{O}(\zeta^2)).$$

  Since we are calculating the "regular type," one of the numbers $a'(0)$, $b'(0)$ must be non-zero. Thus if we let $a(\zeta) = \zeta + \ldots$ then the expression $|z_1^2 - z_2^3|^2$ in the definition of ρ provides the obstruction to the order of contact: the curve cannot have order of contact better than 4. Similar considerations show that if $b(\zeta) = \zeta + \ldots$ then the order of contact cannot exceed 6. Putting these ideas together, we determine that a regular curve that exhibits maximum order of contact at $P = 0$ is $\phi(\zeta) = (0, \zeta, 0)$. Its order of contact with M at P is 6. Thus $\Delta_1^{reg}(M, P) = 6$.

- We may see that $\Delta_1(M, P) = \infty$ by considering the (singular) curve $\phi(\zeta) = (\zeta^3, \zeta^2, 0)$. This curve actually *lies in M*.

Thus we see that type may differ dramatically from regular type. ■

An appealing feature of the notion of analytic finite type in $\mathbb{C}^2$ that we learned about earlier is that it is upper semi-continuous: if, at a point P, the expression $\langle \partial\rho, F\rangle$ is non-vanishing for some $F \in \mathcal{L}_\mu$, then it will certainly be non-vanishing at nearby points. Therefore if P is a point of type m it follows that sufficiently nearby points will be of type at most m. It is considered reasonable that a viable notion of finite type should be upper semi-continuous. Unfortunately, this is not the case as the following example of D'Angelo shows:

EXAMPLE 1.3.12. Consider the hypersurface in $\mathbb{C}^3$ defined by the function

$$\rho(z_1, z_2, z_3) = \operatorname{Re} z_3 + |z_1^2 - z_2 z_3|^2 + |z_2|^4.$$

Take $P = 0$. Then we may argue as in the last example to see that $\Delta_1(M, P) = \Delta_1^{reg}(M, P) = 4$. The curve $\zeta \mapsto (\zeta, \zeta, 0)$ gives best possible order of contact.

But for a point of the form $P = (0, 0, ia)$, a a positive real number, let α be a square root of ia. Then the curve $\zeta \mapsto (\alpha\zeta, \zeta^2, ia)$ shows that $\Delta_1(M, P) = \Delta_1^{reg}(M, P)$ is at least 8 (in fact it equals 8—Exercise).

Thus we see that the number Δ_P is not an upper semi-continuous function of the point P. ■

It has recently been shown by Yu ([**YU1**]) and by McNeal that, on a convex domain in any dimension, the notion of type is truly upper semi-continuous.

D'Angelo [**DAN2**] proves that the invariant Δ_1 can be compared with another invariant that comes from intersection-theoretic considerations; that is, he compares Δ_1 with the dimension of the quotient of the ring $\mathcal{O}$ of germs of holomorphic functions by an ideal generated by the components of a special decomposition of the defining function. This latter *is* semi-continuous. The result gives essentially sharp bounds on how Δ_1 can change as the point P varies within M.

It is worth discussing briefly why singular varieties are needed to detect type in dimensions three and above. Assume that $\Omega \subseteq \mathbb{C}^n$ is a smooth domain and $P \in \partial\Omega$ is normalized so that $P = 0$ and the $2n - 1$ dimensional Euclidean tangent plane to $\partial\Omega$ at P is $\{\operatorname{Re} z_n = 0\}$. If $n = 2$ then it is easy to argue that a complex variety with best order of contact to $\partial\Omega$ at P must have the form $\zeta \mapsto (\phi_1(\zeta), 0), \phi_1(0) = 0$. If it happens that ϕ_1' vanishes to order at least one at 0 then we can divide out a suitable power of ζ and arrange for $\phi_1'(0) \neq 0$, thus desingularizing the variety. Notice that this simple observation is special to dimension $n = 2$. In dimension three, the variety of best contact will have the form $\zeta \mapsto (\phi_1(\zeta), \phi_2(\zeta), 0)$. A typical instance is $\zeta \mapsto (\zeta^2, \zeta^3, 0)$. This could indeed be the variety of best contact (as the examples below show) but it cannot be desingularized.

We give now a brief description of the algebraic invariant that is used in [**DAN2**]. Take $P \in M$ to be the origin. Let ρ be a defining function for M near 0. The first step is to prove that one can write the defining function in the form

$$\rho(z) = 2\mathrm{Re}\, h(z) + \sum_j |f_j(z)|^2 - \sum_j |g_j|^2, \tag{2}$$

where h, f_j, g_j are holomorphic functions.

We now describe the genesis of (2) in the case that ρ is a polynomial. Let ρ be a locally defined polynomial defining function for a hypersurface M in $\mathbb{C}^n$. By standard techniques (see [**KRA1**]), we may write

$$\rho(z) = 2\mathrm{Re}\,(\mu(z)) + \mathcal{M}(z),$$

where $\mathcal{M}$ consists only of terms involving monomials that contain both z's and $\overline{z}$'s. A simple change of variable enables us to then write the hypersurface in the form

$$\rho(z) = 2\mathrm{Re}\, z_n + \hat{\mathcal{M}}(z),$$

where $\hat{\mathcal{M}}$ again is a sum of mixed monomials.

We may think of $\hat{\mathcal{M}}$ as a Hermitian quadratic form on a finite dimensional vector space of holomorphic monomials. That is, if z^α and z^β are two such monomials then their inner product $\langle z^\alpha, z^\beta\rangle$ is defined to be

$$\langle z^\alpha, z^\beta\rangle = \frac{1}{2}\left[\left(\frac{\partial}{\partial z}\right)^\alpha \left(\frac{\partial}{\partial \overline{z}}\right)^\beta \rho(0) + \left(\frac{\partial}{\partial \overline{z}}\right)^\alpha \left(\frac{\partial}{\partial z}\right)^\beta \rho(0)\right].$$

Institute a holomorphic change of coordinates whose Jacobian at the origin diagonalizes this quadratic form. Thus $\hat{\mathcal{M}}$ will have a certain number of positive eigenvalues and a certain number of negative eigenvalues. Letting f_j be eigenfunctions for the former and g_j be eigenfunctions for the latter we find that we have rewritten $\hat{\mathcal{M}}$ as

$$\hat{\mathcal{M}}(z) = \sum_{j=1}^{m} |f_j(z)|^2 - \sum_{j=1}^{m'} |g_j(z)|^2. \tag{3}$$

If $m' > m$ then we set $f_{m+1} = f_{m+2} = f_{m'} \cdots \equiv 0$, so that the vectors $f = (f_1, f_2, \ldots, f_{m'})$ and $g = (g_1, g_2, \ldots, g_{m'})$ take values in the same dimensional space; then we may write our formula more briefly as

$$\hat{\mathcal{M}}(z) = \|f(z)\|^2 - \|g(z)\|^2.$$

A similar device is of course used when $m > m'$. Thus we have established formula (2). [A discussion of the formula (3), from a slightly different point of view, appears in [**DAN4**].]

Since ρ is a polynomial then each sum in (3) can be taken to be finite—say that j ranges from 1 to m in both sums. Let us restrict attention to that case.

[See [**DAN6**] for a thorough treatment of this decomposition and [**KRA5**] for auxiliary discussion.] Write $f = (f_1, \ldots, f_m)$ and $g = (g_1, \ldots, g_m)$.

Let U be a unitary matrix of dimension k. Define $\mathcal{I}(U, P)$ to be the ideal generated by h and $f - Ug$. We set $D(\mathcal{I}(U, P))$ equal to the dimension of $\mathcal{O}/\mathcal{I}(U, P)$. Finally declare $B_1(M, P) = 2 \sup D(\mathcal{I}(U, P))$, where the supremum is taken over all possible unitary matrices of order m. Then we have

THEOREM 1.3.13. *With M, ρ, P as usual we have*

$$\Delta_1(M, P) \leq B_1(M, P) \leq 2(\Delta_1(M, P))^{n-1}.$$

THEOREM 1.3.14. *The quantity $B_1(M, P)$ is upper semi-continuous as a function of P.*

We learn from the two theorems that Δ_1 *is* locally finite in the sense that if it is finite at P then it is finite at nearby points. We also learn by how much it can change. Namely, for points Q near P we have

$$\Delta_1(M, Q) \leq 2(\Delta_1(M, P))^{n-1},$$

In case the hypersurface M is pseudoconvex near P then the estimate can be sharpened. Assume that the Levi form is positive semi-definite near P and has rank q at P. Then we have

$$\Delta_1(M, Q) \leq \frac{(\Delta(M, P))^{n-1-q}}{2^{n-2-q}}.$$

We conclude this section with an informal statement of the theorem of [**CAT2**]:

THEOREM 1.3.15. *Let $\Omega \subseteq \mathbb{C}^n$ be a bounded pseudoconvex domain with smooth boundary. Let $P \in \partial\Omega$. Then the problem $\bar{\partial} u = f$, with f a $\bar{\partial}$- closed $(0, 1)$ form, enjoys a gain in regularity in the Sobolev topology on a neighborhood of P if and only if P is a point of finite type in the sense that $\Delta_1(M, P)$ is finite.*

It is not known how to determine the sharp "gain" in regularity of the $\bar{\partial}$-Neumann problem at a point of finite type in dimensions $n \geq 3$. There is considerable evidence (see [**DAN6**]) that our traditional notion of "gain" as described here will have to be refined in order to formulate a sharp result.

It turns out that to study finite type, and concomitantly gains in Sobolev regularity for the problem $\bar{\partial} u = f$ when f is a $\bar{\partial}$- closed $(0, q)$ form, requires the study of order of contact of q- dimensional varieties with the boundary of the domain. One develops an invariant $\Delta_q(M, P)$. The details of this theory have the same flavor as what has been presented here, but are more technical.

One of the most dramatic applications of the regularity theory of the $\bar{\partial}$- Neumann problem is to the study of boundary regularity of biholomorphic mappings. Let

$$P : L^2(\Omega) \to L^2(\Omega) \cap (\text{holomorphic functions})$$

be the orthogonal projection (this is given by the Bergman kernel—see [**KRA1**]). Bell's Condition R (see [**BEL1**]) is that $P : C^\infty(\bar{\Omega}) \to C^\infty(\bar{\Omega})$. The lesson that we shall now see is that one can verify Condition R using estimates for the $\bar{\partial}$-Neumann problem. We proceed as follows:

In the theory of the $\bar{\partial}$- Neumann problem (see [**FOLK**] and [**KRA7**]) one learns that the Bergman projection P satisfies

$$P = I - \bar{\partial}^* N \bar{\partial}.$$

Here $\bar{\partial}^*$ is the L^2 adjoint of $\bar{\partial}$ and N is the *$\bar{\partial}$-Neumann operator*—a canonical right inverse for the $\bar{\partial}$- Laplacian. Part of proving Theorem 1.3.15 is to see that, when Ω is of finite type, then for each s there is an $\epsilon > 0$ such that $N : W^s(\Omega) \to W^{s+\epsilon}(\Omega)$. But then it is plain (since $\bar{\partial}, \bar{\partial}^*$ are first order partial differential operators) that $P : W^s(\Omega) \to W^{s-2}(\Omega)$. Thus we see that a finite type domain satisfies Condition R. And Bell's theorem is that a biholomorphic map of domains that satisfy Condition R extends smoothly to the boundary. We summarize with a theorem:

THEOREM 1.3.16. *Let Ω_1, Ω_2 be domains of finite type in $\mathbb{C}^n$. If $\Phi : \Omega_1 \to \Omega_2$ is a biholomorphic mapping then Φ extends to a C^∞ diffeomorphism of $\bar{\Omega}_1$ onto $\bar{\Omega}_2$.*

Heartening progress has been made in studying the singularities and mapping properties of the Bergman and Szegö kernels on domains of finite type in both dimension 2 and in higher dimensions. We mention particularly [**NRSW**], [**CHR2**] - [**CHR5**], and [**MCN1**] - [**MCN3**].

There is still a great deal of work to be done before we have reached a good working understanding of points of finite type. One recent development worth noting is that McNeal has shown [**MCN1**] that, for a convex domain in any dimension, the type of a boundary point may be calculated using only non-singular affine complex analytic discs. In other words

$$\text{D'Angelo type} = \text{regular type} = \text{line type}$$

at a convex boundary point. J. Yu [**YU2**] has extended these ideas to multi-type. Thus we see that in the convex case the geometry simplifies considerably.

We conclude by again contrasting the case of dimension $n = 2$ with that of $n > 2$. Why can we measure the geometry using non-singular varieties in dimension two, but not in higher dimensions? Put in other words, why do *singular* varieties not provide more information in dimension two? The answer is almost trivial (and hence not transparent). Examine the form in which we wrote the defining function of a domain in $\mathbb{C}^2$ when we were proving Theorem 1.3.4:

$$\rho(z) = 2\text{Re}\, z_2 + f(z_1) + \mathcal{O}(|z_1 z_2| + |z_2|^2).$$

The type of the point 0, either by the analytic or the geometric definition, was seen to be the order of vanishing of the function $f(z_1)$. What is crucial for us now is that $f(z_1)$ is a function of a single complex variable.

The point is that measuring the order of vanishing of f using singular varieties is no different from measuring that order using non-singular varieties. That is,

$$\frac{\nu(f(\zeta^k))}{\nu(\zeta^k)} = \frac{\nu(f(\zeta^k))}{k} = \nu(f(\zeta)).$$

This is why we may concentrate attention on non-singular curves when we study hypersurfaces in $\mathbb{C}^2$.

As already noted, a more subtle but more profound simplification takes place in *every* dimension when we study a hypersurface that arises as the boundary of a convex domain. In that case it is known (see [**MCN5**], [**YU2**], [**YU3**]) that the geometry may be measured using order of contact of complex lines.

1.4. Aire on Invariant Metrics. The material in the preceding section suggests that we learn some of the basic ideas concerned with the Kobayashi metric and other canonical metrics. The Carathéodory and Kobayashi/Royden metrics are two canonical metrics that have important invariance and extremal properties. For many purposes, these two metrics interact more naturally with function theory than does the Bergman metric. On the other hand, the Bergman metric is a Kähler metric while these last two are only Finsler metrics.

Now let $U \subseteq \mathbb{C}^n, V \subseteq \mathbb{C}^n$ be open sets. We let $U(V)$ denote the set of all holomorphic mappings from V to U. As usual, $B \subseteq \mathbb{C}^n$ is the unit ball.

DEFINITION 1.4.1. If $\Omega \subseteq \mathbb{C}^n$ is open, then the *infinitesimal Carathéodory metric* is given by $F_C : \Omega \times \mathbb{C}^n \to \mathbb{R}$ where

$$F_C(z,\xi) = \sup_{\substack{f \in B(\Omega) \\ f(z)=0}} |f_*(z)\xi| \equiv \sup_{\substack{f \in B(\Omega) \\ f(z)=0}} \left| \sum_{j=1}^n \frac{\partial f}{\partial z_j}(z) \cdot \xi_j \right|.$$

Remark: In this definition, the $\mathbb{C}^n$ in $\Omega \times \mathbb{C}^n$ should be thought of as the tangent space to Ω at z—in other words, $\Omega \times \mathbb{C}^n$ should be canonically identified with the (complexified) tangent bundle of Ω at z. We think of $F_C(z,\xi)$ as the length of the tangent vector ξ at the point $z \in \Omega$. In general, $F_C(z,\xi)$ is not given by a quadratic form $(g_{ij}(z))$, hence F_C is not a Riemannian metric. [In fact, we may not in general hope that the indicatrix of the Carathéodory metric is convex—see [**SIB1**].] It is instead a *Finsler metric.* ■

It is known ([**BSW**] and [**GK2**]) that a generic open set in $\mathbb{C}^n$ that is diffeomorphic to B is not biholomorphic to B. If we recall the proof of the Riemann mapping theorem, we see that the definition of the Carathéodory metric (and of the Kobayashi/Royden metric) takes the extremal problem from that proof ([**AHL1**]) and uses it to measure to what extent the proof fails in general.

Notice that if $f \in B(\Omega), z \in \Omega$, and ϕ is an automorphism of the ball taking $f(z)$ to 0, then $|(\phi \circ f)_*(z)\xi| \geq |f_*(z)\xi|$ (use Schwarz's lemma, for example). Thus the condition $f(z) = 0$ in the last definition is superfluous.

DEFINITION 1.4.2. Let $\Omega \subseteq \mathbb{C}^n$ be open and $\gamma : [0,1] \to \Omega$ a C^1 curve. The *Carathéodory length of a curve* of γ is defined to be

$$L_C(\gamma) = \int_0^1 F_C(\gamma(t), \gamma'(t))\, dt.$$

This definition parallels the definition of the length of a curve in a Riemannian metric. It would be natural at this point to define the (integrated) Carathéodory distance between two points to be the infimum of lengths of all curves connecting them. One advantage of defining distance in this fashion is that it is then straightforward to verify the triangle inequality.

However we shall not take this approach. A crucial feature of the Carathéodory metric is that it is, in a precise sense, the smallest metric under which holomorphic mappings are distance decreasing. The notion of distance suggested in the last paragraph is *not* the smallest. It is fortuitous that the following definition *does* result in the smallest distance-decreasing metric—in particular it *does* satisfy the triangle inequality.

DEFINITION 1.4.3. Let $\Omega \subseteq \mathbb{C}^n$ be an open set and $z, w \in \Omega$. The *Carathéodory distance* between z and w is defined to be

$$C(z,w) = \sup_{f \in B(\Omega)} \rho(f(z), f(w)),$$

where ρ is the Poincaré-Bergman distance on B.

Notice that, depending on the domain Ω, the Carathéodory distance between two distinct points may be 0. (This happens, for instance, if Ω is the complex plane less a set of zero analytic capacity.)

Remark: The Poincaré/Bergman distance $B(z,w)$ between two points z, w in the ball $B \subseteq \mathbb{C}^n$ is given by

$$B(z,w) = \frac{\sqrt{n+1}}{2} \log \left(\frac{|1 - w \cdot \bar{z}| + \sqrt{|w-z|^2 + |z \cdot \bar{w}|^2 - |z|^2|w|^2}}{|1 - w \cdot \bar{z}| - \sqrt{|w-z|^2 + |z \cdot \bar{w}|^2 - |z|^2|w|^2}} \right),$$

where $z \cdot \bar{w} \equiv \sum_j z_j \bar{w}_j$. We leave the details of this assertion as an exercise—see also [**KRA1**]. (*Hint:* First calculate the distance of the origin to a point of the form $(a, 0)$, using the fact that the disc $\{(\zeta, 0, \ldots, 0)\}$ is totally geodesic; then exploit the automorphism group.) Now recall:

DEFINITION 1.4.4. Let $\Omega \subseteq \mathbb{C}^n$ be open. Let $e_1 = (1, 0, \ldots, 0) \in \mathbb{C}^n$. The infinitesimal form of the *Kobayashi/Royden metric* is defined to be the function

$F_K : \Omega \times \mathbb{C}^n \to \mathbb{R}$, where

$$\begin{aligned} F_K(z,\xi) &\equiv \inf\{\alpha : \alpha > 0 \text{ and } \exists f \in \Omega(B) \text{ with } f(0) = z, (f'(0))(e_1) = \xi/\alpha\} \\ &= \inf\left\{\frac{|\xi|}{|(f'(0))(e_1)|} : f \in \Omega(B), (f'(0))(e_1) \right. \\ &\qquad\qquad \left. \text{is a constant multiple of } \xi\right\} \\ &= \frac{|\xi|}{\sup\{|(f'(0))(e_1)| : f \in \Omega(B), (f'(0))(e_1) \text{ is a constant multiple of } \xi\}}. \end{aligned}$$

Here $|\xi|$ denotes Euclidean length.

We now wish to define an integrated distance based on elements of $\Omega(B)$. The natural analogue for our definition of Carathéodory distance does not satisfy a triangle inequality (see [**LEM1**]). Moreover, we want the Kobayashi/Royden distance to be the *greatest* metric under which holomorphic mappings are distance decreasing. Therefore we proceed as follows:

DEFINITION 1.4.5. Let $\Omega \subseteq \mathbb{C}^n$ be open and $\gamma : [0,1] \to \Omega$ a piecewise C^1 curve. The *Kobayashi/Royden length* of γ is defined to be

$$L_K(\gamma) = \int_0^1 F_K(\gamma(t), \gamma'(t))dt.$$

DEFINITION 1.4.6. Let $\Omega \subseteq \mathbb{C}^n$ be open and $z, w \in \Omega$. The *Kobayashi/Royden distance* between z and w is defined to be

$$K(z,w) = \inf\{L_K(\gamma) : \gamma \text{ is a piecewise } C^1 \text{ curve connecting } z \text{ and } w\}.$$

Recall that we did not implement a definition like this one for the (integrated) Carathéodory distance because we were able to find a *smaller* distance that satisfied the triangle inequality. Now we are at the other end of the spectrum: we want the Kobayashi metric to be as large as possible, and to satisfy a triangle inequality as well. This definition serves that dual purpose.

We remark in passing that the use of the ball as a model domain when defining the Carathéodory and Kobayashi/Royden metrics is important, but it is not the unique choice. The theory is equally successful if either the disc or the polydisc is used. However, in the current state of the theory, it is essential that the model domain have a transitive automorphism group. Further ideas concerning the use of "model domains" appear, for instance, in [**GK5**] and [**MA1**].

PROPOSITION 1.4.7 (THE DISTANCE DECREASING PROPERTIES). *If Ω_1, Ω_2 are domains in $\mathbb{C}^n$, $z, w \in \Omega_1$, $\xi \in \mathbb{C}^n$, and if $f : \Omega_1 \to \Omega_2$ is holomorphic, then*

$$\begin{aligned} F_C^{\Omega_1}(z,\xi) \ge F_C^{\Omega_2}(f(z), f_*(z)\xi) \qquad & F_K^{\Omega_1}(z,\xi) \ge F_K^{\Omega_2}(f(z), f_*(z)\xi) \\ C_{\Omega_1}(z,w) \ge C_{\Omega_2}(f(z), f(w)) \qquad & K_{\Omega_1}(z,w) \ge K_{\Omega_2}(f(z), f(w)). \end{aligned}$$

PROOF. This follows by inspection from the definitions. ∎

COROLLARY 1.4.8. *If $f : \Omega_1 \to \Omega_2$ is biholomorphic then f is an isometry in both the Carathéodory and the Kobayashi/Royden metrics.*

PROOF. Apply the proposition to both f and f^{-1}. ■

COROLLARY 1.4.9. *If $\Omega_1 \subseteq \Omega_2 \subseteq \mathbb{C}^n$ then for any $z, w \in \Omega_1$, any $\xi \in \mathbb{C}^n$, we have*

$$F_C^{\Omega_1}(z,\xi) \geq F_C^{\Omega_2}(z,\xi) \qquad F_K^{\Omega_1}(z,\xi) \geq F_K^{\Omega_2}(z,\xi)$$
$$C_{\Omega_1}(z,w) \geq C_{\Omega_2}(z,w) \qquad K_{\Omega_1}(z,w) \geq K_{\Omega_2}(z,w).$$

PROOF. Apply the proposition to the inclusion mapping $i : \Omega_1 \to \Omega_2$. ■
Up to a constant multiple, the Bergman, Carathéodory, and Kobayashi/Royden metrics are equal on the ball B. This follows from the transitivity of the automorphism group on the unit sphere bundle in the complexified tangent bundle.

LEMMA 1.4.10. *Let $B \subseteq \mathbb{C}^n$ be the unit ball and $D_1 \subseteq B$ be the disc $\{(\zeta, 0, \ldots, 0) \in B\}$. Fix $\xi = (t, 0, \ldots, 0) \in \mathbb{C}^n, z \in D_1$. Then*

$$\begin{aligned} F_C^B(z,\xi) &= F_C^{D_1}(z,\xi) \\ F_K^B(z,\xi) &= F_K^{D_1}(z,\xi). \end{aligned}$$

PROOF. Since $D_1 \subseteq B$, Corollary 1.2.9 implies the inequality $\leq$ in both cases. For $\geq$, apply the distance decreasing property to the mapping $\pi_1 : B \to D_1$ given by $(z_1, \ldots, z_n) \mapsto (z_1, 0, \ldots, 0)$. ■
Here is an alternative definition for the integrated Kobayashi/ Royden metric that is particularly useful on complex manifolds. For domains in $\mathbb{C}^n$ it is not difficult to see that this new definition is equivalent to the one given above [exercise]. The definitions are also equivalent on manifolds (see [**ROY**]), but this is rather tricky to prove.

DEFINITION 1.4.11. Let $\Omega \subseteq \mathbb{C}^n$ and fix $z, w \in \Omega$. Call a set $\{p_0, \ldots, p_k\}$ *admissible* for $\{z, w\}$ if $p_0 = z, p_k = w$, and there exist $f_j \in \Omega(B)$ and points $u_j^1, u_j^2 \in B, j = 1, \ldots, k$, with $f_j(u_j^1) = p_{j-1}, f_j(u_j^2) = p_j$.

Set

$$K(z,w) = \inf_{\substack{\{p_0,\ldots,p_k\} \\ \text{admissible}}} \sum_{j=1}^{k} B(u_j^1, u_j^2),$$

where B is the Bergman distance on the ball B.

PROPOSITION 1.4.12. *Let $\Omega \subseteq \mathbb{C}^n$ be an open set. Let d be a metric on Ω that satisfies $d(f(z), f(w)) \leq B(z,w)$ for all $f \in \Omega(B), z, w \in B$ (here $B(z,w)$ is the Poincaré-Bergman distance). Then $d(z,w) \leq K(z,w)$.*

PROOF. Let $z, w \in \Omega, f_1, \ldots, f_k, p_0, \ldots, p_k, u_1^1, \ldots, u_k^1, u_1^2, \ldots, u_k^2$ be as in the second definition of the integrated Kobayashi/Royden metric (using admissible sequences). Then

$$\begin{aligned} d(z,w) &\leq \sum_{j=1}^{k} d(p_{j-1},p_j) \\ &= \sum_{j=1}^{k} d(f_j(u_j^1), f_j(u_j^2)) \\ &\leq \sum_{j=1}^{k} B(u_j^1, u_j^2). \end{aligned}$$

Passing to the infimum on the right hand side gives

$$d(z,w) \leq K(z,w),$$

as desired. ∎

PROPOSITION 1.4.13. *Let $\Omega \subseteq \mathbb{C}^n$ be an open set. Let d be any metric on Ω that satisfies $d(z,w) \geq B(f(z), f(w))$ for all $f \in B(\Omega)$ and $z, w \in \Omega$. Then $d(z,w) \geq C_\Omega(z,w)$.*

PROOF. This is a formality. ∎

PROPOSITION 1.4.14. *Let $\Omega \subseteq \mathbb{C}^n$ be open. Then $K_\Omega(z,w) \geq C_\Omega(z,w)$ for all $z, w \in \Omega$.*

PROOF. Use one of the preceding two propositions. ∎

In the function theory of one complex variable, the Poincaré metric can be transplanted from the disc to any planar domain by way of the uniformization theorem (see [**FIS**] or [**FAK**]). Thus one does not generally see the Carathéodory or Kobayashi metrics in one variable treatments (see, however, [**KRA9**]).

Precious little is known about calculating the Bergman, Carathéodory, or Kobayashi/Royden metrics for general domains. One can use the uniformization theorem to obtain some fairly sharp information about smoothly bounded domains in $\mathbb{C}^1$. No such tool is available in several variables.

For special domains such as the ball, or bounded symmetric domains, the automorphism group is a powerful tool for obtaining explicit formulas. Fefferman obtained information about the asymptotic boundary behavior of the Bergman metric on a strongly pseudoconvex domain in [**FEF**]. No theorem of this sort has ever been proved on any more general classical group of domains—although see some encouraging generalizations by Gebelt [**GEB**] and Diederich [**DIH**]. Some explicit calculations of the Kobayashi metric, on domains without a transitive group of automorphisms, is done in [**BFKKMP**].

For the Carathéodory and Kobayashi metrics one of the most striking results is due to I. Graham [**GRA1**]. We now describe it:

THEOREM 1.4.15 (I. GRAHAM). *Let $\Omega \subset\subset \mathbb{C}^n$ be a strongly pseudoconvex domain with C^2 boundary. Fix $P \in \partial\Omega$. Let $\xi \in \mathbb{C}^n$, and write $\xi = \xi_T + \xi_N$, the* *decomposition of ξ into complex tangential and normal components relative to the geometry at the point P.*

Let ρ be a defining function for Ω normalized so that $|\nabla\rho(P)| = 1$. Let $\Gamma_\alpha(P)$ be a non-tangential approach region at P. If F represents either the Carathéodory or Kobayashi/Royden metric on Ω then

$$\lim_{z\to P} d_\Omega(z)\cdot F(z,\xi) = \frac{1}{2}|\xi_N|.$$

Here $|\ |$ denotes Euclidean length and $d_\Omega(x)$ is the distance of x to the boundary of Ω.

If $\xi = \xi_T$ is complex tangential, then we have

$$\lim_{\Gamma_\alpha(P)\ni z\to P} \sqrt{d_\Omega(z)}\cdot F(z,\xi) = \frac{1}{2}\mathcal{L}(\xi,\xi). \tag{1}$$

Here $\mathcal{L}$ is the Levi form calculated with respect to the defining function ρ.

Graham's result is striking for several reasons. First, it gives a direct connection between the interior geometry of Ω (vis à vis the Carathéodory and Kobayashi/Royden metrics) and the boundary geometry (vis à vis the Levi form). Second, it provides enough information to see that both of these metrics are *complete* [Exercise]. Some generalizations and refinements of Graham's theorem appear in [**ALA1**]. In particular, Aladro found a sharp form of Graham's theorem that treats both normal and tangential directions simultaneously. He achieves this result by inserting weights before the (decomposed) tangent vector ξ on the left hand side of (1) and performing delicate calculations.

Graham's and Aladro's results show us, in particular, that the $\delta\times\sqrt{\delta}$ geometry of Stein's theorem is just right for strongly pseudoconvex domains. It turns out that this statement is true *only* for strongly pseudoconvex domains.

We should mention here that it would be of interest to be able to interpret the Levi geometry of a weakly pseudoconvex boundary point using interior geometry, such as the Kobayashi metric. Important progess on this program has been made by J. Yu [**YU4**].

The method of proof of Graham's theorem is an important one: he exploits, in a precise way, the fact that a strongly pseudoconvex point is "very nearly" like a boundary point of the unit ball. We shall not present the details of this argument.

Sibony [**SIB2**] has constructed an interesting invariant metric that always lies between the Carathéodory and Kobayashi metrics:

DEFINITION 1.4.16. Let Ω be a domain in $\mathbb{C}^n$ (or more generally a complex manifold). If $P\in\Omega$ we let $\mathcal{S}_P$ denote the class of functions $u(z)$ on Ω that are C^2 in a neighborhood of P and satisfy $u(P) = 0, 0\le u\le 1$, and $\log u$ is plurisubharmonic on all of Ω. If $\xi\in\mathbb{C}^n$ and $\mathcal{L}$ represents the complex Hessian operator then we define

$$S_\Omega(P,\xi) = \sup_{u\in\mathcal{S}_P} \left(< \mathcal{L}u(P)\xi,\xi >\right)^{1/2}.$$

This metric has many nice properties: it is homogeneous and subadditive in the second entry; it renders holomorphic mappings distance decreasing, and it always lies between the Carathéodory and Kobayashi metrics. But it is much easier to localize and to estimate than these other two, simply because plurisubharmonic functions are much more flexible objects than are holomorphic functions. Sibony constructed his metric to study problems about hyperbolicity of manifolds.

More recently, Lempert [**LEM1**] has obtained sharp results about the Kobayashi metric on strongly *convex* domains. Call a domain $\Omega \subset\subset \mathbb{C}^n$ with C^2 boundary *strongly convex* if all of its boundary curvatures are positive (i.e. the real Hessian of a defining function is strictly positive definite).

THEOREM 1.4.17 (LEMPERT [**LEM1**]). *Let $\Omega \subset\subset \mathbb{C}^n$ be strongly convex with C^6 boundary. Fix $P \in \Omega$. For each $\xi \in \mathbb{C}^n, |\xi| = 1$, there is a uniquely determined function $\phi_\xi : D \to \Omega$ such that*

a: *$\phi'_\xi(0)$ is a scalar multiple of ξ;*

b: *$\phi_\xi(0) = P$;*

c: *$F_K(P,\xi) = 1/|\phi'_\xi(0)|$.*

If $\zeta \in B, \zeta \neq 0$, then we may write ζ uniquely in the form $\zeta = r\xi$, where $0 < r < 1$ and $\xi \in \partial B$. Define a mapping $F : B \to \Omega$ by

$$F(\zeta) = \begin{cases} \phi_\xi(r) & \text{if } \zeta \neq 0 \\ P & \text{if } \zeta = 0. \end{cases}$$

Then F extends to be a C^2 diffeomorphism of $\bar{B} \setminus \{0\}$ to $\bar{\Omega} \setminus \{P\}$.

This is the only known result, for a general class of domains, about the global behavior of extremal discs for the Kobayashi metric. Sibony [**SIB1**] has shown that Lempert's results fail categorically on general strongly pseudoconvex domains. [Another approach to some of Lempert's ideas, using a dual extremal problem, appears in [**ROW**].] It remains a mystery to see what Lempert's results might mean (or even suggest) for a class of domains that is indigenous to the main stream of several complex variables.

A crucial lemma in Lempert's work states that the integrated Kobayashi/Royden metric on a convex domain can be realized by a Kobayashi/Royden chain (Definition 1.2.11) of length one. I would like to see an analogue of Lempert's ideas developed for strictly pseudoconvex domains, but with chains of length supplanted by chains of some fixed length. Crucial to the success of such a program would be the following:

CONJECTURE: Let $\Omega \subseteq \mathbb{C}^n$ be a strongly pseudoconvex domain. Then there is a positive integer K with the property that the integrated Kobayashi/Royden distance between any two points $P, Q \in \Omega$ can be realized with a Kobayashi/Royden chain of length at most K.

It should be noted that M. Y. Pang ([**PAN1**], [**PAN2**]) has found both necessary and sufficient conditions for this conjecture to hold. They are all related to the convexity of the indicatrix for the Kobayashi metric. These results suggest that it might prove useful to study the convexification of the Kobayashi metric. We also refer to recent results of Venturini [**VEN**] that bear on this problem.

In a recent private communication, J. P. Rosay has given an example of a smoothly bounded, pseudoconvex domain on which the the Carathéodory and Kobayashi metrics are not comparable; more precisely, the infinitesimal Carathéodory metric is bounded on the domain, while the infinitesimal Kobayashi metric is not. By contrast the two metrics (as well as the Bergman metric—see Section 2.4) *are* comparable on strongly pseudoconvex domains and on finite type domains (see Section 1.3) in $\mathbb{C}^2$. There is clearly much yet to be understood about these metrics.

1.5. Boundary Limits of Holomorphic Functions. Fatou's striking theorem, which appeared in 1906, just a few years after Lebesgue published the foundations of his measure theory, is as follows:

THEOREM 1.5.1 (FATOU, 1906 [**FAT**]). *Assume f is a bounded holomorphic function on the unit disc $D \subseteq \mathbb{C}$. Then for almost every $\theta \in [0, 2\pi)$ it holds that*

$$\lim_{r \to 1^-} f(re^{i\theta}) \equiv f^*(e^{i\theta})$$

exists.

Fatou's proof is an interesting mirror of the ideas that were in the air at the time. He thought of f as

$$f(re^{i\theta}) = \sum_{j=0}^{\infty} a_j r^j e^{ij\theta}.$$

For a fixed θ, claiming that $\lim_{r \to 1^-} f(re^{i\theta})$ exists is just the same thing as claiming that the series $\sum_{j=0}^{\infty} a_j e^{ij\theta}$ is Abel summable. But Abel summability is in turn implied by Césaro summability. So in fact Fatou used Fourier series techniques to see that the series is Césaro summable for almost every θ.

While of definite analytic interest, Fatou's proof did not begin to reveal the geometry of theorems of this kind. This geometry began to emerge when Privalov-Plessner proved their extension of Fatou's theorem:

For $\alpha > 1$ and $\theta \in [0, 2\pi)$ we define

$$\Gamma_\alpha(e^{i\theta}) = \{z \in D : |z - e^{i\theta}| < \alpha(1 - |z|)\}.$$

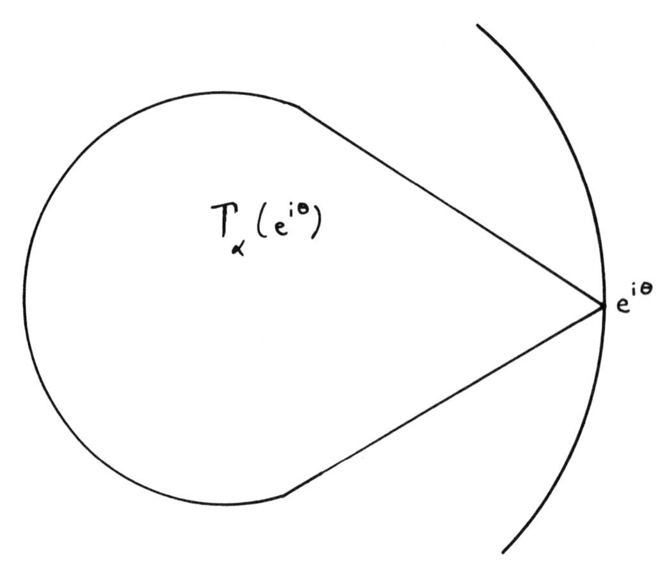

FIGURE 1.12

The region $\Gamma_\alpha(e^{i\theta})$ is depicted in Figure 1.12. It is called a *non-tangential approach region* or, more classically, a "Stolz angle." Now we have

THEOREM 1.5.2. *Let f be a bounded holomorphic function on the disc $D \subseteq \mathbb{C}$. Fix $\alpha > 1$. Then for almost every $\theta \in [0, 2\pi)$ it holds that*

$$\lim_{\Gamma_\alpha(e^{i\theta}) \ni z \to e^{i\theta}} f(z) = f^*(e^{i\theta}).$$

Central to the proof of this theorem is an estimate on the Poisson kernel for the disc. This kernel may be written as

$$P_r(e^{i\theta}) = \frac{1}{2\pi} \frac{1 - r^2}{1 - 2r\cos\theta + r^2}.$$

Fix $\alpha > 1$. There is a $C = C_\alpha > 0$ such that if $z = re^{i\psi} \in \Gamma_\alpha(e^{i\theta})$ then

$$0 \leq P_r(e^{i\psi}) \leq C_\alpha \cdot P_r(e^{i\theta}). \tag{1}$$

It is therefore plausible that the non-tangential boundary behavior of a (reasonable) holomorphic function is controlled by its radial boundary behavior. See [**KRA1**] for more on these matters.

But there is more here than meets the eye. First, the Lindelöf principle teaches us that the phenomenon of radial behavior controlling non-tangential behavior occurs *pointwise* for bounded analytic functions—not just almost everywhere. A version of the result is this:

THEOREM 1.5.3 (LINDELÖF). *Let f be a bounded holomorphic function on the disc $D \subseteq \mathbb{C}$. Fix $\theta \in [0, 2\pi)$. Suppose that*

$$\lim_{r \to 1^-} f(re^{i\theta}) \equiv \ell$$

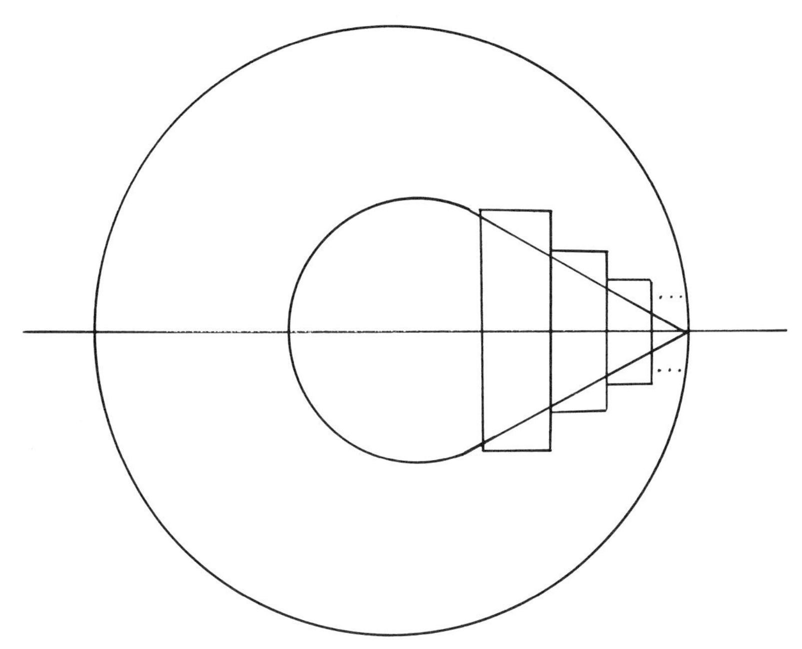

FIGURE 1.13

exists. Then for any $\alpha > 1$ *it holds that*

$$\lim_{\Gamma_\alpha(re^{i\theta})\ni z\to e^{i\theta}} f(z) = \ell.$$

SKETCH OF PROOF: For simplicity assume that $\theta = 0$. Cover $\Gamma_\alpha(1)$ with boxes, each of which is conformally equivalent to the first one under a complex affine mapping that preserves the real axis. See Figure 1.13. Let the conformal mapping of B_1 to B_k be called ϕ_k. Consider the functions $\{f \circ \phi_k\}$. Since f is bounded, these form a normal family. Thus there is a normally convergent subsequence $\{f \circ \phi_{k_j}\}$. Since this subsequence must converge to ℓ on the real axis, it in fact must converge normally to ℓ on all of B_1 (by the argument principle). But, unraveling the definitions, this means that f tends to ℓ non-tangentially as z tends to 1. ■

Lehto and Virtanen [**LEV**] made two profound observations about the Lindelöf principle. One is that a weaker condition than boundedness of f suffices to guarantee that $\{f \circ \phi_k\}$ is a normal family. The second is that the relationship among the boxes B_k (they can all be thought of as images, under *conformal self-maps of the entire disc*, of B_1) suggests that hyperbolic geometry is at work. They were led to the following definition:

DEFINITION 1.5.4. Let f be a meromorphic function on the disc $D \subseteq \mathbb{C}$. We say that f is a *normal* function on D if whenever $\{\psi_k\}$ is a family of conformal self-maps of the disc then $\{f \circ \psi_k\}$ is a normal family.

To repeat what we said above, we can now mildly rework the original proof of the Lindelöf principle, noting that it can be arranged not only for the boxes to be linearly conformally equivalent, but also that they be equivalent *under Möbius transformations of the entire disc.* To see this in detail requires some calculations; but the key insight is that an approach region $\Gamma_\alpha(1)$ is comparable to a region of the form

$$\mathcal{U}_\alpha \equiv \{z \in D : d_P(z, I) < C_\alpha\}, \tag{2}$$

where I is the segment $\{x + i0 : 0 < x < 1\}$ and d_P represents the Poincaré, or hyperbolic, distance in D. Thus, in sum, the Lindelöf principle holds for normal functions.

To see that there are normal functions other than bounded ones, notice that any holomorphic function on the disc that omits two complex values, or any meromorphic function that omits three, is normal. Indeed one may compose with the elliptic modular function to reduce this assertion to the case of a bounded function. It follows that a Schlicht function is normal since (by topology) its image must omit a continuum. There is a large lore of normal functions in one complex variable—see, for instance, [**ACP**]. We shall discuss normal functions in greater detail in Section 2.2.

As we have noted, Lehto and Virtanen also saw the importance of hyperbolic geometry in the Lindelöf phenomenon. This insight enabled them to prove the following strengthening of the theorem:

THEOREM 1.5.5. *Let $f(z)$ be a normal function in the disc D. Assume that f has an asymptotic boundary limit α along a Jordan curve, the curve lying in the closure of D, that terminates at the boundary point P. Then f has non-tangential limit α at P.*

Efforts have been made to extend the theory of normal functions to the context of several complex variables, but there have been some surprises. Let us begin by examining the ball in $\mathbb{C}^2$. If we follow our instincts then it is natural to suppose that the right approach regions to use when studying the boundary behavior of holomorphic functions would be regions $\mathcal{U}_\alpha$ defined as in line (2) above. Instead of the Poincaré metric one could use the Poincaré-Bergman metric, or the Carathéodory metric, or the Kobayashi/Royden metric: on the ball they are all comparable (indeed the same up to a normalizing constant). After a calculation, one sees that for the domain $\Omega = B$ the approach region $\mathcal{U}_\alpha$ is comparable to the *admissible approach region* of Koranyi/Stein that is defined, for $P \in \partial B$, by

$$\mathcal{A}_\alpha(P) = \big\{z \in B : |1 - \langle z, P\rangle| < C_\alpha \cdot (1 - |z|)\big\}.$$

It has the feature that in complex normal directions it is non-tangential in shape while in complex tangential directions it is parabolic in shape. (A special case of) Koranyi's theorem [**KOR1**], [**KOR2**] is:

THEOREM 1.5.6. *Let f be a bounded holomorphic function on B. Let $\alpha > 1$. Then for almost every boundary point $P \in \partial B$ it holds that*

$$\lim_{\mathcal{A}_\alpha(P) \ni z \to P} f(z) \equiv f^*(P)$$

exists.

This being the case, one might suppose that a Lindelöf principle obtains for bounded analytic functions on the ball, with non-tangential approach regions replaced by admissible approach regions. Unfortunately such a result is false, as was first discovered by Cirka [**CIR**]. Consider the following example:

EXAMPLE 1.5.7. Define the holomorphic function

$$f(z_1, z_2) = \frac{{z_2}^2}{1 - z_1}$$

on B. Because

$$|z_2|^2 < 1 - |z_1|^2 = (1 - |z_1|)(1 + |z_1|) \leq 2|1 - z_1|,$$

we see that f is bounded on B. Moreover,

$$\lim_{r \to 1^-} f((r, 0)) = 0.$$

The sequence $w^j = (1 - 1/j, 1/\sqrt{j})$ approaches the boundary point $(1, 0)$ admissibly through the approach region $\mathcal{A}_2((1, 0))$. However

$$\lim_{j \to \infty} f(w^j) = 1.$$

Thus f does not have an admissible limit at $(1, 0)$. ■

The *proof* of the classical Lindelöf principle breaks down in the multi-variable context because the radial segment $I = \{(x + i0, 0) : 0 < x < 1\}$ is no longer a set of uniqueness for holomorphic functions. Thus there is a loss of information. The following more restrictive result is essentially the best that one can do:

THEOREM 1.5.8 (CIRKA, CIMA/KRANTZ). *Let $P \in \partial B \subseteq \mathbb{C}^2$ and let f be a bounded holomorphic function on B. Fix $P \in \partial B$. Suppose that*

$$\lim_{r \to 1^-} f(rP) = \ell.$$

If $w^j = (u^j, v^j)$ is any sequence in B that satisfies (for some $\alpha > 1$)

$$\frac{|1 - w^j \cdot \bar{P}|}{1 - |w^j|} < \alpha$$

and

$$\limsup_{j \to \infty} \frac{|w^j - (w^j \cdot \bar{P})P|^2}{|1 - w^j \cdot \bar{P}|} = 0 \tag{3}$$

then

$$\lim_{j \to \infty} f(w^j) = \ell.$$

The hypothesis (3) about the sequence $\{w^j\}$ in the theorem guarantees that the sequence approaches P through a region that is asymptotically smaller, in the complex tangential directions, than every admissible approach region. It is more convenient to express this hypothesis geometrically: it states that

$$\lim_{j\to\infty} d(w^j, I_P) = 0,$$

where I is the segment stretching from the origin to P and d is the distance in the hyperbolic (Poincaré-Bergman) metric.

It turns out that the right way to think about normal functions, both in one and several complex variables, is as follows: Let $\Omega \subseteq \mathbb{C}^n$ be a bounded domain. A holomorphic function $f : \Omega \to \mathbb{C}$ is normal if and only if its derivative is bounded from the Kobayashi metric on Ω to the spherical metric on $\mathbb{C} \subseteq \Sigma$, where Σ is the Riemann sphere. With this point of view, results like those enunciated in the last paragraph become rather natural. The concept of normal function has been generalized to functions taking values in a complex manifold in [**ALA2**]. We shall examine normal functions in greater detail in Section 2.2.

We will, in the next section, outline an approach to Fatou theorems and to Lindelöf theorems that will use invariant (hyperbolic) geometry to unify the two phenomena, both in one and several complex variables. This approach was inspired by the normal functions considerations presented here. As a side benefit, we will recapture the Hartogs extension phenomenon as an instance of the boundary behavior of holomorphic functions. It also provides a device for discovering boundary limit phenomena along lower dimensional manifolds, such as those discovered in [**NARU**].

1.6. A Geometric Look at the Fatou Theorem. Before beginning, we wish to stress the following fact: the point of view being developed here could not be gleaned, nor even suspected, from considering just one complex variable, or just the ball, or just pseudoconvex domains. Rather, one must consider all domains at once, and in all dimensions. One seeks to understand how the boundary (Levi) geometry of a smoothly bounded domain determines the function theory on the interior. Notice that this program is essentially vacuous in one complex variable. For if $\Omega \subseteq \mathbb{C}$ is any smoothly bounded domain and if U is a small, topologically trivial boundary neighborhood (see Figure 1.14), then $\Omega \cap U$ is conformally equivalent to the disc.

Moreover, classical results (see [**BEK**] for an exposition) guarantee that this conformal map is smooth to the boundary portion $U \cap \partial\Omega$ as illustrated in the figure (essentially as smooth as the boundary will admit). Thus the theory of boundary behavior of a holomorphic function on Ω—with attention restricted to $U \cap \partial\Omega$—is really no different from the theory of boundary behavior on the disc.

Matters are quite different in two or more complex variables. For in that context, generically, any two domains are biholomorphically inequivalent—even locally. Let us make that claim more substantive. Let $\Omega_0 \subseteq \mathbb{C}^n$ be the unit ball.

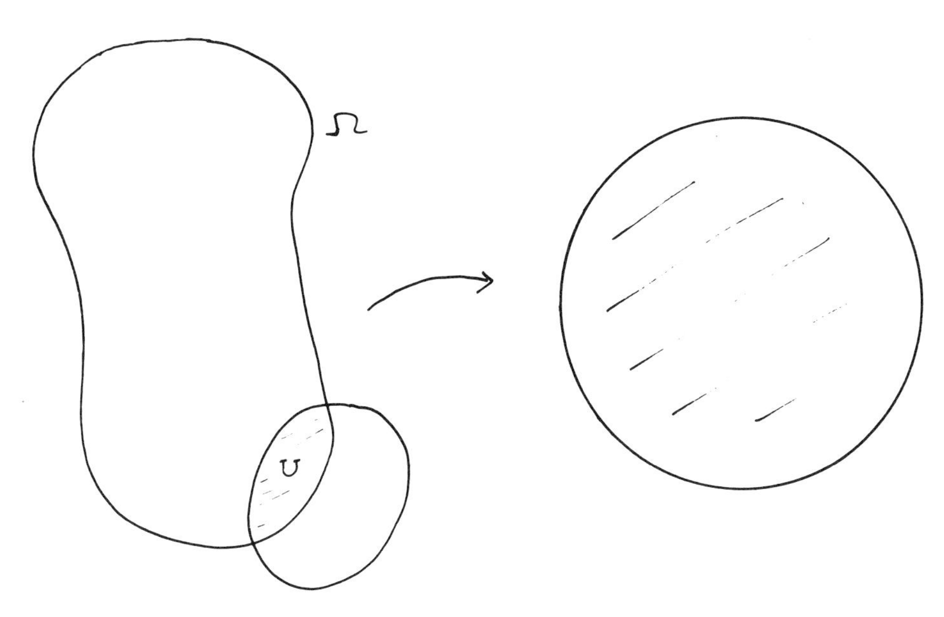

FIGURE 1.14

It is convenient to think of Ω_0 as the sublevel set of a *defining function*:

$$\Omega_0 = \{z \in \mathbb{C}^n : \rho_0(z) \equiv \|z\|^2 - 1 < 0\}.$$

If ρ is any defining function then set $\Omega_\rho = \{z \in \mathbb{C}^n : \rho(z) < 0\}$. To avoid aberrations, we shall always assume that $|\nabla \rho| = 1$ on $\partial\Omega$. For $\epsilon > 0$ we define the collection

$$\mathcal{O} = \mathcal{O}_\epsilon^X \equiv \{\Omega_\rho : \|\rho - \rho_0\|_X < \epsilon\}.$$

Here X is a function space norm to be specified. For instance X could be C^1 or C^k or C^∞ or a suitable topology on C^ω. *Notice that $\mathcal{O}$ is a collection of domains.* We define an equivalence relation on $\mathcal{O}$ by $\Omega_1 \sim \Omega_2$ if and only if Ω_1 is biholomorphic to Ω_2. How many equivalence classes are there?

It is known that, no matter what the topology on X (chosen from among the ones listed), and as long as $0 < \epsilon < 1/10$, say, then in dimension one there is just one equivalence class. This is just the Riemann mapping theorem. However in dimensions two and higher the number of equivalence classes is uncountable. Moreover, the phenomenon is a local one. If one sets up equivalence classes of nearby pieces of hypersurface, the answer is the same. So the *ansatz* of "any piece of boundary is locally like the boundary of the disc, therefore ... " cannot work in several variables. [A good reference for this discussion is [**GK1**], in which further primary literature is cited.]

In spite of the differences that we have suggested between the theories of one and several complex variables, it would be most attractive to have a theory of the boundary behavior of holomorphic functions of one and several variables that treats both situations simultaneously and reveals—for essentially intrinsic reasons— why the one variable situation is somewhat homogeneous and the several variable situation rather diverse. We now sketch such a theory using the Kobayashi metric. First, we provide a sort of genesis of the ideas.

As we have sketched in the one variable case, the classical approach to boundary behavior of bounded holomorphic functions, or more generally to functions in Hardy classes, was motivated by specific kernels. The nature of the singularity of the Poisson kernel on the disc determined the shape of the non-tangential approach regions. An inequality of the form (1) in the last section could not hold on a region that is essentially broader than a non-tangential region. (However see [**NAS1**], [**SUE1**],[**SUE2**],[**SUE3**] for some startling variants. It turns out that what is essential is not so much the shape of the approach region as the area of its cross sections.) Likewise, the shape of the singularity of the Poisson kernel in *several real variables* suggests the use of approach regions that are non-tangential (or conical) in shape. In particular, the following estimate for the Poisson kernel $P : \Omega \times \partial\Omega \to \mathbb{R}$ proves to be crucial in shaping the nature of the theorems.

PROPOSITION 1.6.1. *If $\Omega \subset\subset \mathbb{R}^N$ is a domain with C^2 boundary, then*

$$P_\Omega(x, y) \approx \frac{\delta_\Omega(x)}{|x - y|^N}.$$

The proof of this result is elementary but elaborate—the technique of comparison domains, and the maximum principle, are the main tools. The only source in print for the proofs is [**KRA1**], in which the proofs of Norberto Kerzman are reproduced.

In the course of proving estimates of the form (1) in Section 5 one needs to work on regions in which the numerator of the Poisson kernel controls (a power of) the denominator. This is the motivation for the definition of the classical non-tangential approach regions

$$\Gamma_\alpha(y) = \{x \in \Omega : |x - y| < \alpha \cdot \delta_\Omega(x)\}.$$

The standard theorem is that if h is a bounded harmonic function on a smoothly bounded domain in $\mathbb{R}^N$ then h has almost-everywhere boundary limits through non-tangential approach regions. Of course there are analogous results for functions with p-mean growth, $1 < p < \infty$.

Koranyi [**KOR1**],[**KOR2**] discovered in 1969 (and this was generalized by Stein [**STE3**] in 1972) how to exploit the singularity of a less well-known canonical kernel to see further than the Poisson kernel allowed. At first, let us restrict attention to the ball in $\mathbb{C}^n$. Koranyi considered the *Poisson-Szegö* kernel

$$\mathcal{P}(z, \zeta) = c_n \frac{(1 - |z|^2)^n}{|1 - \langle z, \zeta\rangle|^{2n}}$$

on the ball. Notice that this kernel is positive; it reproduces holomorphic functions on the ball that are continuous on the closure; and it has certain formal similarities to the classical Poisson kernel (for a detailed treatment of this kernel see [**KRA7**] as well as the representation theory literature). The singularity of this kernel is a power of $|1 - \langle z, \zeta \rangle|$, which is the same expression that is used to define the admissible approach regions $\mathcal{A}_\alpha$. Formally, then, one might hope that the same majorization techniques that led to the Privalov-Plessner theorem might lead to a theorem on the ball about admissible limits of holomorphic functions. That is the thrust of [**KOR1**],[**KOR2**].

We note in passing that, while the Poisson-Szegö kernel is not widely known in analysis circles, it is certainly a standard artifact of constructions in representation theory and in partial differential equations.

Stein's contribution, which for us is a milestone, is to free these considerations from kernels. For domains that lack a large automorphism group, there is little hope of actually calculating the canonical kernels (although there has been heartening progress on strongly pseudoconvex, convex, finite type, and a few other domains—see [**FEF**], [**BOS**], [**MCN1**],[**MCN2**],[**MCN3**], [**KRA1**] and references therein).

Now fix a smoothly bounded domain $\Omega \subset\subset \mathbb{C}^n$, any $n \geq 1$. [Much of what we are about to say is valid in principle on domains with Lipschitz or rougher boundary, but a more unified treatment is achieved at this point if we mandate at least C^2 boundary.] What is special about our new point of view is that we begin *not* with a particular kernel to determine the relevant geometry (as in Privalov/Plessner [**PRI**] or Koranyi [**KOR1**],[**KOR2**]), and *not* with a predisposed idea of parabolic geometry (as in Stein [**STE3**]), and we do not restrict to a special case where certain calculations are possible (as in [**NSW1**], [**NSW2**], [**NRSW**], although these papers represent the most important special case of the ideas presented here). The point of view is instead to begin with the intrinsic geometry and build the function theory on that foundation. Details of this construction can be found in [**KRA8**].

We equip Ω with the Kobayashi/Royden metric. It is elementary to see that, on a bounded domain, this is a genuine metric (that is, the domain is *hyperbolic*). It is an open problem to determine whether the metric is complete, even when the domain is pseudoconvex and of finite type (however see [**CAT4**] for important results in dimension two). The example of [**FOK**], discussed in Section 1 of the present chapter, suggests that completeness may not always hold, even for a smoothly bounded pseudoconvex domain. Fix a number $\alpha > 0$ once and for all. All of the subsequent constructs depend on this number α. It plays a role analogous to the aperture α of non-tangential approach regions in the classical theory and of the aperture α of admissible approach regions in the Koranyi/Stein theory.

Let U be a tubular neighborhood of $\partial\Omega$. Choose $\epsilon > 0$ such that if $P \in \partial\Omega$ then $\bar{B}(P, 2\epsilon) \subseteq U$. [Here B denotes a Euclidean ball.] To each such P, we associate the segment $i_P \equiv \{P - t\nu_P : 0 \leq t \leq \epsilon\}$. Here ν_P is the unit outward normal vector to $\partial\Omega$ at P. The set i_P is the *segment* with base at P that points in the normal direction at P and emanates *into* the domain. In what follows, we let $d = d_K$ denote *distance in the Kobayashi/Royden metric.*

Set

$$\mathcal{U}_\alpha(P) = \{z \in \Omega : d(z, i_P) < \alpha\}.$$

Here $d(z, i_P) \equiv \inf_{w \in I_P} d(z, w)$. Then the sets $\mathcal{U}_\alpha(P)$ will play the role of approach regions for studying the boundary behavior of holomorphic functions. Some comments about this definition are in order. First, the elementary proof of the Lindelöf principle that results from this definition (see [**CIK**]) is a part of the motivation. Also, any definition formulated in terms of the Kobayashi/Royden metric is more obviously invariant than the *ad hoc* definition given in either [**STE3**] or [**KRA1**]. It has recently become apparent, through the work of Huang [**HUA3**] and [**MER**], that we may replace i_P in the definition of $\mathcal{U}_\alpha(P)$ by a *complex geodesic terminating at* P. This is the best of all possible situations, for then the definition of $\mathcal{U}_\alpha(P)$ is *prima facie* invariant under biholomorphic mappings. It is something of a nuisance to verify this invariance using the more classical definitions (see [**STE3**]).

We let π denote (Euclidean) normal projection to the boundary of Ω. We will define a family of balls in the boundary of Ω by

$$\beta(P, r) \equiv \pi\left\{ z \in \mathcal{U}_\alpha(P) : \mathrm{dis}_{Eucl}(z, \partial\Omega) = r \right\}.$$

Here π is the standard Euclidean orthogonal projection to the boundary.

Using the theory of spaces of homogeneous type as motivation (see [**COW1**], [**COW2**] and Sections 2.3, 2.6), we realize that a successful harmonic analysis on the boundary $\partial\Omega$ of our domain requires both a suitable family of balls and a compatible *measure*. Departing from other treatments, we construct the measure from the pre-existing balls.

In most instances the Kobayashi metric has associated to it a volume form V. This can be defined in a number of different fashions. One way to do this is as follows: Let $\Omega(B)$ denote the holomorphic mappings of the unit ball B into Ω. For $P \in \Omega$, let the *Kobayashi volume form*

$$M_K^\Omega(P) = \inf\{1/|\det f'(0)| : f \in \Omega(B), f(0) = P\}.$$

Then we let

$$dV = M_K^\Omega \, d\bar{z}_1 \wedge dz_1 \wedge \cdots \wedge d\bar{z}_n \wedge dz_n.$$

An alternative construction, which in many contexts is equivalent to this one, is due to Eisenman [**EIS**]. It uses a variant of the Carathéodory measure construction (as in the standard construction of Hausdorff measure—see [**FED**]) to construct a Kobayashi volume element.

Now we specify the sense in which the complex function theory should be consistent with the geometry provided by the metric being used:

DEFINITION 1.6.2. Fix a smoothly bounded domain $\Omega \subseteq \mathbb{C}^n$. A pair (ρ, V), consisting of a metric ρ on Ω and a volume V on Ω, is said to be *compatible* with Ω if, for every metric ball $B(Z,\alpha)$ in Ω, it holds that $0 < V(B(Z,\alpha)) < \infty$ and there is a constant C such that, for all u plurisubharmonic on Ω, it holds that

$$|u(Z)| \leq C \cdot \frac{1}{V(B(Z,\alpha))} \int_{B(Z,\alpha)} |u(\zeta)|\, dV(\zeta) .$$

We now construct measures on $\partial\Omega$ using the balls $\beta(P,r)$ and the volume V. Our construction is a variant of the Carathéodory measure construction as found in [**FED**]. It will yield a family of measures that detect analytic objects of varying dimensions

We think of the domain Ω, a compatible pair (ρ, V), and an $\alpha > 0$ as all being fixed. A *gauge function* ω is a function

$$\omega : \{x \in \mathbb{R} : x \geq 0\} \to \{x \in \mathbb{R} : x \geq 0\}$$

which is continuous and nondecreasing. Fix such an ω. Let $\delta > 0$. Let $E \subseteq \partial\Omega$ be a set. A countable collection of balls $\{\beta(P_i, r_i)\}_{i=1}^{\infty}$ is called a *δ-admissible* cover for E if $0 < r_i \leq \delta$ for all i and

$$\bigcup_i \beta(P_i, r_i) \supseteq E .$$

Let $A_\delta(E)$ be the collection of all δ-admissible covers of E.

Define the outer measure

$$\mu_\delta(E) = \inf \left\{ \sum_i \omega(r_i) \cdot V(B(P_i - r_i \nu_{P_i}, \alpha)) : \{\beta(P_i, r_i)\} \in A_\delta(E) \right\} .$$

Observe that if $0 < \delta_1 < \delta_2$ then $\mu_{\delta_1}(E) \geq \mu_{\delta_2}(E)$. Therefore we may define

$$\mu(E) = \lim_{\delta \to 0^+} \mu_\delta(E) .$$

Then μ is the outer measure which we shall use to formulate the Fatou theorem. The presence of the terms $V(B(P_i - r_i\nu_{P_i}, \alpha))$ in the definition of μ has a practical role, as will be seen in the proof of the theorems, but it also has a philosophical role. The latter is that the condition of compatibility is invariant under scaling of V. Since we will estimate the maximal function over $\mathcal{U}_\alpha$ in terms of boundary maximal functions involving the measure μ, and since the device for mediating between the two is the compatibility, it is necessary to build V into the definition of μ.

In practice, particularly in the examples below, we will choose $\omega(r)$ to be a power of r. The particular power chosen will be determined by (1) the boundary geometry of the domain Ω, (2) the dimension of the ambient space $\mathbb{C}^n$, and (3) the dimension of the analytic object in $\partial\Omega$ that is being studied.

We mandate that the balls $\beta(\zeta, r)$ not vary wildly. When this mild hypothesis is satisfied we call the metric "tame." It is essentially automatic for the

Kobayashi metric on a smoothly bounded domain. The concept of "tame" is described more precisely as follows:

A metric space is called "directionally limited" if, in effect, the number of points that can lie on a sphere of radius $r > 0$ and that are bounded from each other by r is *a priori* bounded above. The full technical definition is in [**FED**], p. 145. Finite-dimensional normed vector spaces, Riemannian manifolds, and locally Euclidean spaces are the prime examples of directionally limited metric spaces.

We define a binary operation on $\partial\Omega$ by

$$\tau(a,b) = \min\left\{\inf\{r : a \in \beta(b,r)\}, \inf\{r : b \in \beta(a,r)\}\right\} .$$

In case $\partial\Omega$ (or an open subset U of $\partial\Omega$) is a directionally limited metric space when equipped with the binary function τ, then we call the original metric ρ on Ω *tame.*

We will assume in practice that if $\beta = \beta(\zeta, r)$ is a boundary ball and $B = B(\zeta - r\nu_\zeta, \alpha)$ the metric ball in Ω of which it is the projection then

$$V(B) \cdot \omega(r) \leq C \cdot \mu(\beta) . \qquad (P)$$

We refer to this condition as property (P). In all practical applications in which the Caratheéodory construction is successful (see [**FED**]), this type of estimate holds. The purpose of the Caratheéodory construction is to take a scalar valued function on certain distinguished sets and turn it into a measure; however, one usually wants the measure of the distinguished sets to be comparable to the original scalar valued function on those sets.

We also will assume in practice that the measure $d\mu$ is absolutely continuous with respect to area measure $d\sigma$ on $\partial\Omega$. Note that this hypothesis, together with property (P), implies that the μ-measure of a boundary ball $\beta(\zeta, r)$ will always be positive and finite. When we formulate the principal result of the paper, we shall further discuss the significance of the hypothesis that $d\mu$ be absolutely continuous with respect to $d\sigma$.

If μ is constructed as above and if $f \in L^1(d\mu)$ then we define

$$M_2 f(P) = \sup_{0<r<\epsilon_0} \frac{1}{\mu(\beta(P,r))} \int_{\beta(P,r)} |f(\zeta)| \, d\mu(\zeta) .$$

We reserve the notation $M_1 f(P)$ to denote the traditional Hardy–Littlewood maximal function defined using area measure $d\sigma$ and the isotropic balls b: We set

$$M_1 f(P) = \sup_{0<r<\epsilon_0} \frac{1}{\sigma(b(P,r))} \int_{b(P,r)} |f(\zeta)| \, d\sigma(\zeta)$$

whenever $f \in L^1(d\sigma)$.

Using the language developed in the last several paragraphs, we can now formulate our theorem. Let $\Omega \subseteq \mathbb{C}^n$ be a bounded domain with C^2 boundary. Let Ω be equipped with a metric ρ and a volume form V such that (ρ, V) is compatible with Ω. Assume that the metric ρ is tame. Fix $\alpha > 0$ and define approach regions $\mathcal{U}_\alpha$ as in Section 4. Fix a gauge function ω and define the corresponding measure $d\mu$. Let $U \subseteq \partial\Omega$ be open and assume that $d\mu$ is absolutely continuous with respect to $d\sigma$ and has property (P) on U.

THEOREM 1.6.3. *If $f \in H^p(\Omega)$, $0 < p \leq \infty$, then for μ-almost every $P \in U$ it holds that*

$$\lim_{\mathcal{U}_\alpha(P) \ni z \to P} f(z) = f^*(P) .$$

Next we give a number of examples illustrating the significance of the theorem.

The approach sketched here for studying the boundary behavior of holomorphic functions, and more generally the function theory of a domain in $\mathbb{C}^n$, is not in final form. It is not completely satisfactory in that it requires the verification of several properties of the measure and the balls. We suspect that these properties in fact hold virtually all the time, but current techniques are inadequate to verify them. Nonetheless, the following examples provide ample reason to suppose that this is a worthwhile method for studying the complex analysis of domains We shall see that the theorem not only subsumes the known results about boundary behavior of H^p functions, but it reveals some new ones.

EXAMPLE 1.6.4. Let $\Omega = D$ be the unit disc in $\mathbb{C}^1$. Let ρ be the Poincaré metric

$$\|\xi\|^2_{\rho,Z} = \frac{2|\xi|^2}{(1-|Z|^2)^2}$$

and let dV be given by

$$dV = \frac{2}{(1-|z|^2)^2}(-i)\, d\bar{z} \wedge dz .$$

Recall (see [**GAR**]) that a metric ball $B(Z, \alpha)$ equals a classical Euclidean disc with center

$$\frac{1-a^2}{1-a^2|Z|^2} Z$$

and Euclidean radius

$$a\frac{1-|Z|^2}{1-a^2|Z|^2} ,$$

where

$$a = \frac{e^\alpha - 1}{e^\alpha + 1} .$$

Once α is fixed, then one sees that this disc has Euclidean center comparable to Z and Euclidean radius comparable to $(1 - |Z|)$. Therefore

$$V\big(B(Z, \alpha)\big) \approx C ,$$

for some constant C (depending on α but independent of Z). As a result, compatibility reduces to the usual sub-mean value property for subharmonic functions on Euclidean discs.

Recall that the classical nontangential or Stolz regions on the disc are defined by

$$\Gamma_\alpha \equiv \{z \in D : |z - P| < \alpha \cdot (1 - |z|)\}, \quad 1 < \alpha < \infty .$$

Since the Poincaré metric is essentially Euclidean distance scaled by distance to the boundary, it is easy to check that there are positive constants $A(\alpha)$ and $B(\alpha)$ such that the approach regions $\mathcal{U}_\alpha(P)$ are comparable to these classical regions in the sense that

$$\Gamma_{A(\alpha)}(P) \subseteq \mathcal{U}_\alpha(P) \subseteq \Gamma_{B(\alpha)}(P) .$$

It follows that the balls $\beta(P, r)$ in the boundary are arcs centered at P of radius comparable to r. (Note that the comparability constants depend on α, as does the whole theory.)

Because of this last fact about the balls β, we use the gauge function $\omega(r) = r$. It follows that the resulting measure μ is (mutually absolutely continuous with respect to) one-dimensional Hausdorff measure, or arc length measure. Thus $d\mu$ is absolutely continuous with respect to $d\sigma$ and has property (P). In conclusion, Theorem 1.6.3 is nothing other than the classical nontangential convergence theorem of Privalov, Plessner. ■

EXAMPLE 1.6.5. Let $\Omega = B$, the unit ball in $\mathbb{C}^n$. Let ρ be any of the Bergman, Caratheéodory, Kobayashi, or Sibony metrics (they are identical, up to normalizing constants), given by

$$\|\xi\|^2_{\rho,z} \quad = \quad \sum \bar{\xi}_i \frac{n+1}{(1-|z|)^2} \left[(1-|z|^2)\delta_{ij} + \bar{z}_i z_j\right] \xi_j . \tag{1}$$

Let

$$dV = (1 - |z|^2)^{-n-1}(-i)^n d\bar{z}_i \, d\bar{z}_1 \wedge dz_1 \wedge \cdots \wedge d\bar{z}_n \wedge dz_n .$$

This is the usual volume associated with, say, the Bergman metric.

Fix $\alpha > 0$. Let us examine the metric balls $B(Z, \alpha)$. Without loss of generality we confine our attention to

$$P = \mathbb{1} = (1, 0) \in \partial B \subseteq \mathbb{C}^2 .$$

Since the disc $\mathbf{d} = \{(\zeta, 0) : \zeta \in D\}$ is a totally geodesic submanifold of B in the Bergman metric and since the restriction of the metric (1) to $\mathbf{d}$ is the Poincaré metric on this disc, we may take advantage of the calculations in Example 1.6.4. We find that for a complex normal vector $\xi = (\xi_1, 0)$ at a point $Z = (t, 0)$ near $\mathbb{1}$, $0 < t < 1$, the metric has the form

$$\|\xi\|_{\rho,Z} = \frac{\sqrt{3}}{(1 - |Z|^2)} |\xi| .$$

Next note that for the same Z and the complex tangent vector $\xi = (0, \xi_2)$ the metric (1) reduces to

$$\|\xi\|_{\rho,Z} = \frac{\sqrt{3}}{(1 - |Z|^2)^{1/2}} |\xi| .$$

Taking $t = 1 - \delta$ we find that the metric ball $B(Z, \alpha)$ is comparable to the bidisc

$$D(1 - \delta, \delta/2) \times D(0, \sqrt{\delta}/2) ,$$

with the constants of comparability depending on α. As a result, $V(B(Z, \alpha))$ is essentially constant, with the constant depending only on α (*not* on Z). The compatibility of (ρ, V) now follows from the sub-mean value property of a plurisubharmonic function on a bidisc.

From the shape of the balls $B(Z, \alpha)$ we conclude that the metric approach regions $\mathcal{U}_\alpha(\mathbb{1})$ are comparable to the admissible approach regions of Koranyi (see [**KOR1**],[**KOR2**]). As a result, the boundary balls $\beta(\mathbb{1}, r)$ are comparable to the nonisotropic balls of Koranyi.

We choose the gauge function $\omega(r) = r^n$ (because this is, up to scaling, the $(2n-1)$-dimensional Euclidean area measure of a boundary Koranyi ball of radius r). The result of this choice of ω and the essential constancy of the volumes of metric balls inside Ω is that the corresponding boundary measure μ is (mutually absolutely continuous with respect to) $(2n-1)$-dimensional Hausdorff measure on $\partial\Omega$. Thus the hypotheses of Theorem 1.6.3 are fulfilled. In conclusion, Theorem 1.6.3 in this context reduces to the theorem of Koranyi in [**KOR1**]. ∎

EXAMPLE 1.6.6. Let $\Omega \subseteq \mathbb{C}^n$ be a strongly pseudoconvex domain with C^2 boundary. According to deep work of Graham [**GRA1**] and Fefferman [**FEF**], the Bergman, Caratheéodory, and Kobayashi metrics all have qualitatively the same asymptotic boundary behavior as these metrics do on the unit ball. It is easy to calculate, or follows from the work of Graham, that the Sibony metric does likewise. Thus all of the estimates in Example 1.6.5 apply in the present context. The result is that Theorem 1.6.3, *when applied to a strongly pseudoconvex domain*, reproduces the results of Stein [**STE3**]. Recall that Stein proved his theorems for *any* bounded domain with C^2 boundary, but his theorem is only best possible in the strongly pseudoconvex case. See the next example for a discussion of how Stein's full result, for any domain, can be derived from our Theorem 1.6.3. ∎

EXAMPLE 1.6.7. Let $\Omega \subseteq \mathbb{C}^n$ be *any* bounded domain with C^2 boundary. Let $\delta(z) = \text{dist}(z, \partial\Omega)$, where "dist" denotes Euclidean distance. Let $\varphi(z)$ be a function that is C^∞ on the closure of Ω, supported in U_{ϵ_0} (a tubular neighborhood of $\partial\Omega$ of radius ϵ_0), and identically 1 in a neighborhood of $\partial\Omega$. Define a Hermitian metric

$$\rho = \Sigma \rho_{ij} dz_i d\bar{z}_j ,$$

where

$$\rho_{ij}(z) = \frac{\partial^2}{\partial z_i \partial \bar{z}_j} \Big(C|z|^2 + \varphi(z) \cdot \log\big(1/\text{dist}(z, \partial\Omega)\big) \Big)$$

and C is a large positive constant. This is the same metric that Stein uses in the second half of his book [**STE3**] when he is studying the Lusin area integral on strongly pseudoconvex domains. It has the property that if v is a complex normal vector then

$$\|v\|_{\rho,z} \approx \delta(z)^{-1}|v|$$

while if v is a complex tangential vector then

$$\|v\|_{\rho,z} \approx \delta(z)^{-1/2}|v| \ .$$

See [**KLE**] where metrics of this kind are studied extensively.

The associated volume form is

$$V(z) = \delta(z)^{-n-1}(-i)^n \, d\bar{z}_1 \wedge dz_1 \wedge \cdots \wedge d\bar{z}_n \wedge dz_n \ .$$

The metric balls $B(Z, \alpha)$ are, as shown by Stein ([**STE3**]), comparable to polydiscs as in the last two examples. The compatibility of (ρ, V) now follows because a plurisubharmonic function satisfies the sub-mean value property on a polydisc. It also follows that the quantity $V(B(Z, \alpha))$ is essentially constant (independent of P), with the constants of comparison depending only on α and Ω.

Of course the metric approach regions $\mathcal{U}_\alpha$ now end up being the classical "admissible" approach regions of Stein as in [**STE3**], with parabolic shape in complex tangential directions. After projecting, we see that the balls β in $\partial\Omega$ are the skew balls, with parabolic eccentricity, which are defined by Stein in [**STE3**]. If we take our gauge function to be $\omega(r) = r^n$ then the resulting boundary measure is (mutually absolutely continuous with respect to) $(2n-1)$-dimensional area measure on $\partial\Omega$.

We conclude that Theorem 1.6.3 for this noncanonical metric on an arbitrary domain recovers the theorem of Stein in [**STE3**]. However we stress that this result is optimal *only* in the strongly pseudoconvex case. The next example makes this last point clearer. ■

EXAMPLE 1.6.8. Let $\Omega \subseteq \mathbb{C}^2$ be a domain of finite type in the sense of Kohn [**KON1**] or Bloom/Graham [**BLG**]. A good—if overly simple—example to keep in mind is a domain $E(m_1, m_2)$ of the form

$$|z_1|^{2m_1} + |z_2|^{2m_2} < 1 \ ,$$

where the m_j are positive integers. We take ρ to be any of the Bergman, Caratheéodory, Kobayashi, or Sibony metrics. Catlin [**CAT4**] has shown that the first three metrics are comparable on any domain of finite type in $\mathbb{C}^2$ and it follows from [**SIB2**] that the fourth metric is also.

Fix $\alpha > 0$ and let $P \in \partial\Omega$. Aladro's calculations in [**ALA1**] show that the approach regions $\mathcal{U}_\alpha(P)$ are comparable to the approach regions calculated by Nagel/Stein/Wainger in [**NSW1**],[**NSW2**]. His calculations also show that the polydiscs constructed in [**NSW1**] are comparable to the metric balls $B(Z, \alpha)$ (see also [**CAT4**]). Thus the verification of compatibility reduces to the usual sub-mean value property for $|f|^p$ on polydiscs when f is holomorphic.

Rather than detail these rather technical results here, we point out for example that on the domain $E = E(m_1, m_2)$ (with $m_1, m_2 > 1$) at the boundary point

$e_1 = (1,0)$ (resp., $e_2 = (0,1)$) the approach region is nontangential in complex normal directions and is flat to order $2m_2$ (resp., $2m_1$) in complex tangential directions. At strongly pseudoconvex points $P \in \partial E$ the approach regions are, by Example 1.6.6, parabolic in the tangential directions; however, the parabolicity degenerates as either boundary point $e_1 ore_2$ is approached.

The balls $\beta(P,r)$ in the boundary of a finite type domain are comparable to those constructed by Nagel/Stein/Wainger in [**NSW1**],[**NSW2**]. This too follows from Aladro's calculations.

For the boundary measure, it is convenient to observe that the strongly pseudoconvex points form a dense open set in the boundary of a finite type domain in $\mathbb{C}^2$ (in fact work of Catlin [**CAT2**] implies that in any dimension the strongly pseudoconvex points form an open set of full $d\sigma$-measure in the boundary). Thus we interpret Theorem 1.6.3 by taking $U \subseteq \partial\Omega$ to be the interior of a compact set of *strongly pseudoconvex points* and, following the cue of Example 1.6.6, take the gauge function $\omega(r)$ to be r^2. The resulting measure is, in effect, Euclidean area measure on U *with a weight* (for it tends to zero near any weakly pseudoconvex point). But it is mutually absolutely continuous with respect to area measure on U.

In summary, Theorem 1.6.3 recovers the results of Nagel/Stein/Wainger on a large class of domains of finite type in $\mathbb{C}^2$.

In order to obtain the full result of Nagel/Stein/Wainger, one may exploit the results of [**NSW2**] to ascertain that the balls β satisfy the axioms of K. T. Smith. From this, and delicate calculations, one may verify that $d\mu$ is both absolutely continuous with respect to $d\sigma$ and satisfies property (P) on the entire boundary of a finite type domain in $\mathbb{C}^2$. ∎

EXAMPLE 1.6.9. Let $\Omega \subseteq \mathbb{C}^2$ be the bidisc. Note that the boundary of this Ω is not smooth. However we will only discuss boundary behavior near $(1,0) \in \Omega$, and the result of Theorem 1.6.3 is local.

Let

$$U = \{(e^{i\theta}, z_2) : |\theta| < \pi/2, |z_2| < 1/2\} .$$

Then U is an open set of points in the topological boundary of Ω that is disjoint from the distinguished boundary. What does Theorem 1.6.3 tell us about the boundary behavior of holomorphic functions at such points?

Let ρ be the Kobayashi metric on Ω and V the usual associated volume (see [**EIS**]). For the point $Z = (1-\delta, 0)$ the Kobayashi length of a normal vector $\xi = (\xi_1, 0)$ is determined by the maps

$$D \xrightarrow{\varphi} \Omega \xrightarrow{\pi_1} D$$

where D is the unit disc and

$$\varphi(\zeta) = (1 - \delta + \delta\zeta, 0), \qquad \pi_1(z_1, z_2) = z_1 .$$

It follows that

$$\|\xi\|_{\rho,Z} \approx \frac{|\xi|}{\delta} .$$

On the other hand, if $\xi = (0, \xi_2)$ is a complex tangent vector then the maps

$$D \xrightarrow{\psi} \Omega \xrightarrow{\pi_2} D ,$$

where $\psi(\zeta) = (1-\delta, \zeta)$ and $\pi_2(z_1, z_2) = z_2$, show that

$$\|\xi\|_{\rho,Z} \approx |\xi| .$$

Notice that this last result gives a size for complex tangent vectors that is independent of δ.

It is now easy to see that the metric ball $B(Z, \alpha)$ equals the bidisc

$$D(1-\delta, c(\alpha)\delta) \times D(0, C(\alpha)) ,$$

where c and C are constants that depend on α. The approach region of aperture α at $P = (1,0) \in \partial\Omega$ consequently has the form

$$\mathcal{U}_\alpha(P) = \{z_1 : |z_1 - 1| < K(\alpha)(1 - |z_1|)\} \times \{z_2 : |z_2| < C(\alpha)\} .$$

In short, $\mathcal{U}_\alpha(P)$ is the product of a nontangential approach region in the complex normal direction with a disc in the tangential direction. Therefore a boundary ball $\beta(P_r)$ is the product of a segment of length $k(\alpha) \cdot r$ in the complex normal direction with the same disc $D(0, C(\alpha))$ in the tangential direction. Note that all these calculations also apply, with a small change in the constants, to the other points of any set $U \subseteq \partial\Omega$ that is bounded away from the distinguished boundary.

The only possible choice for ω is $\omega(r) = r$. Then the boundary measure μ turns out to have the property that $\mu(E)$ measures, up to a scaling constant, the diameter of E in the $\operatorname{Im} z_1$ direction. Such a measure is certainly not absolutely continuous with respect to $d\sigma$ (it does have property (P), however). However *ad hoc* arguments show that its role in Theorem 1.6.3 is secure.

Theorem 1.6.3 tells us that on a set of full μ measure in U a function in $H^p(\Omega)$ has boundary limit through a region that is nontangential in normal directions and unrestricted in complex tangential directions. This apparently startling result is in fact nothing more than one gets from

1: the fact that radial limits exist at almost every point of U by Stein's theory and

2: applying Cauchy estimates on the analytic discs $\zeta \to (1-\delta, \zeta)$.

Unfortunately the measure μ picks out at most one point from each of the boundary discs $(e^{i\theta}) \times D(0, 1/2)$. So Theorem 1.6.3 gives the right approach regions but yields a measure that is too thin.

However, one can remedy this situation by exploiting the parameter α as follows. The constant $C(\alpha)$, which determines the radius of the tangential discs, tends to 0 with α. If one constructs μ using not just the balls $\beta(P, r)$ for *fixed* α

but instead the balls $\beta(P, r)$ with $\alpha = r$ and takes $\omega(r) = r$ then the resulting measure is (mutually absolutely continuous with respect to) three-dimensional area measure on U. It should be noted that when one alters the measure construction in this way then the trade-off is that the theorem only applies to the *intersection* of the approach regions $\mathcal{U}_\alpha(P)$—in other words one recovers only radial boundary approach. ■

EXAMPLE 1.6.10. Let Ω be the domain

$$\{(z_1, z_2) \in \mathbb{C}^2 : |z| < 2, |z - (2,0)| > 1\} .$$

Note that the boundary of this Ω is not smooth. However we will only discuss boundary behavior near $P = (1,0) \in \Omega$, and the result of Theorem 1.6.3 is local.

We consider

$$U = \partial\Omega \cap \{z : |z - (1,0)| < 1/2\} .$$

Let ρ be any of the Bergman, Caratheéodory, or Kobayashi metrics on Ω and V the usual associated volume (see, for instance, [**EIS**] or [**KRA1**]). Easy calculations with analytic discs as in the last example show that if $P \in U$ and $Z \in \Omega$ then the ρ-distance of Z to P is finite (see [**KRA3**] for more on the behavior of the Kobayashi metric near such boundary points). In fact the integrated distance will be comparable to Euclidean distance. It follows that an approach region $\mathcal{U}_\alpha(P)$ will always contain a set of the form $\{z \in \Omega : |z - P| < c_0\}$. Thus approach at P will be unrestricted in all directions. Likewise the boundary balls $\beta(P, r)$ will be (comparable to) Euclidean balls $b(P, r)$. The compatibility condition thus follows near U because it holds for Euclidean balls.

Because the balls β are essentially Euclidean, we are forced to choose the gauge $\omega(r) = r^{2n-1}$. This gives a measure that is (mutually absolutely continuous with respect to) $(2n-1)$-dimensional area measure on U. Thus Theorem 1.6.3 says that functions in H^p have boundary limits in the unrestricted sense at μ-almost every point of U. This is the Hartogs extension phenomenon or Kugelsatz in the category of H^p space theory: holomorphic functions on Ω will continue analytically past $(1,0) \in \partial\Omega$ to a fixed neighborhood of $(1,0)$. The H^p space theory ought to reflect that fact, and because of Theorem 1.6.3 it does.

In fact, as J. Polking pointed out to us, one would like to have the H^p theory say that functions in H^p have unrestricted limits at *every* point of U. In Example 1.6.13, where we consider lower-dimensional phenomena, we are able to prove this fact also. ■

EXAMPLE 1.6.11. Fix a domain Ω and consider the Euclidean metric ρ and Euclidean volume V. Fix $\alpha > 0$. Of course (ρ, V) is compatible with Ω *on balls* $B(P, \alpha)$ *which are entirely contained in* Ω. However if $\text{dist}(P, \partial\Omega) << \alpha$ then the compatibility fails in general. So the theory presented in this paper does not (usually) apply for the Euclidean metric.

The one notable exception to this last statement is that, near a strongly pseudoconcave point (as in the last example), compatibility can be verified for the Euclidean metric and volume (this is another manifestation of the Kugelsatz). Thus our theory does apply and Example 1.6.10 is recovered in slightly different language. ■

EXAMPLE 1.6.12. Let Ω be the convex domain

$$\left\{(z_1, z_2) : |z_1|^2 + 2 \cdot \exp\left[-\frac{1}{|z_2|^2}\right] < 1\right\} .$$

This domain is strongly pseudoconvex except at the points $S = \{(e^{i\theta}, 0)\}$. At those special points, the Kobayashi metric in the tangential and normal directions can be estimated either by scaling arguments or by comparison techniques. The metric ball at a point $(1-\delta, 0)$ has size comparable to δ in the normal complex direction and tangential size comparable to $[\log(1/\delta)]^{1/2}$. Of course the metric balls at the strongly pseudoconvex points have the customary product structure, degenerating as S is approached, as in Example 1.6.8. We note that McMichael [**MCM**] has (essentially) done the calculation necessary to check the boundary becomes a directionally limited metric space under the binary function τ. Thus verifying compatibility reduces to the usual sub-mean value property for holomorphic functions on polydiscs. The choice of gauge function $\omega(r) = r\log(1/r)$ gives rise to a measure satisfying our usual hypotheses on $\partial\Omega \setminus S$, and Theorem 1.6.3 applies. ■

EXAMPLE 1.6.13. We restrict attention (for convenience only) to the unit ball B in $\mathbb{C}^2$ equipped with the Kobayashi metric and corresponding volume V. We use the gauge function $\omega(r) = r$ to construct the measure μ. The result is a measure analogous to one-dimensional Hausdorff measure but which assigns Euclidean one-dimensional length to complex normal curves and zero length to complex tangential curves. The reason for this is as follows: consider a boundary curve $\gamma : [0, a] \to \partial\Omega$ of Euclidean length λ. If the curve is complex normal then, for $\delta > 0$, the curve can be covered by approximately λ/δ balls $\beta(P, \delta)$ (since said balls have radius comparable to δ in the complex normal direction). Since each such ball has ω-gauge δ, the approximation to the μ-measure of γ is essentially λ. However if the curve γ is complex tangential then the curve can be covered by $\ll \lambda/r$ balls (one would like to say by $\lambda/\sqrt{r}$ balls, but that is not quite accurate) and the approximations to the μ-measure converge to zero.

Now that this measure has been constructed, we may imitate the proof of our main theorem *on a fixed complex normal curve* γ using the one dimensional measure μ. Observe that we do not attempt to do analysis on the entire boundary at once, but rather on the one-dimensional curve γ. Everything goes through as before with a few important caveats:

(1) One cannot expect the boundary limit of an $H^p(\Omega)$ function along a one-dimensional curve to be in L^p of that curve. More specifically, if that curve is the boundary of a transversal analytic disc—say the disc $\mathbf{d} = \{(\zeta, 0) : |\zeta| < 1\} \subseteq B \subseteq$

$\mathbb{C}^2$—then one cannot expect the restriction of an $f \in H^p(\Omega)$ to lie in $H^p(\mathbf{d})$. Thus we can prove nothing about boundary limits of $H^p(\Omega)$ functions along boundary curves. According to the theory that has already been established by Nagel/Rudin [**NARU**], Ramey [**RAM1**],[**RAM2**],[**RAM3**], Cima/Krantz [**CIK**], and others, this was to be expected.

(2) We might expect a result for $H^\infty(\Omega)$ because the restriction of a bounded function to a lower-dimensional set is still bounded.

(3) The hypothesis that $d\mu$ be absolutely continuous with respect to $d\sigma$ and have property (P) is replaced by the fact (not hypothesis) that in this one-dimensional setting $d\mu$ is absolutely continuous with respect to one-dimensional Hausdorff measure on the curve γ. The argument as to why this implies that the classical Hardy–Littlewood operator is bounded with respect to the measure $d\mu$ on the curve then proceeds via a Poisson operator for a second-order uniformly elliptic partial differential operator on a smooth, transverse two-dimensional surface with γ in its boundary.

(4) It was discovered by Nagel/Rudin [**NARU**] and independently by the present author (unpublished) that in fact the pointwise convergence that one obtains for H^∞ functions along complex normal curves — indeed even for the Lindelöf principle at a single point (see [**CIR**], [**CIK**], and [**KRA1**]) — is not the full admissible convergence (in the strongly pseudoconvex case) or $\mathcal{U}$-convergence (in more general cases). In fact the convergence takes place only through so-called hypoadmissible regions which are asymptotically smaller than the admissible or $\mathcal{U}$-regions (see the condition imposed in line (3) of Section 1.5). A precise rendition of the concept of "asymptotically smaller" is given in [**CIK**] or [**KRA1**]. Until now, there has been no satisfactory explanation for why the convergence on lower-dimensional sets is worse than on sets of positive measure. For instance, in the case of the Lindelöf Principle in one complex variable the hypothesis of radial limit for a bounded analytic function at a point gives rise to a nontangential limit at that point — *not* a limit through an asymptotically smaller region. Why are matters worse in several variables? The geometric method presented here gives an answer to this question: Still restricting attention to Ω being the unit ball B in $\mathbb{C}^2$, let $\gamma(t) = (e^{it}, 0)$. An H^∞ function f on B will have limits almost everywhere along γ in the hypoadmissible sense just described, but not in the full admissible sense. In particular, the points $a_\delta = (1-\delta, \sqrt{\delta})$ have the property that $\lim_{\delta\to 0^+} f(e^{i\theta}a_\delta)$ will in general not agree with $\lim_{\delta\to 0^+} f(e^{i\theta}(1-\delta), 0)$ on a set of points $(e^{i\theta}, 0)$ of positive one-dimensional measure in γ. (Refer also to Section 1.5 for a pointwise version of this failure.) The reason for this negative result, from the point of view of our one-dimensional measure μ constructed above, is as follows. Let $\mathcal{U}_\alpha(\mathbb{1})$ be an approach region with base point $\mathbb{1}$ and with $\alpha > 0$ chosen so that the tangential radius of the approach region at Euclidean distance δ from the boundary is essentially $\sqrt{\delta}$. In particular the approach region should be selected so that it just contains the points a_δ. Then metric balls β centered at $\pi(a_\delta)$ will certainly

intersect the curve γ, but only barely. In particular, the intersection will be a segment of length *much smaller than the anticipated length* δ (because the balls $\beta(a_\delta, r)$ taper at their ends). Therefore the control that the non-isotropic maximal function usually provides at points near the edge of the approach region (for these lower-dimensional phenomena) is *less* than the control provided for points near the center of the approach region. Thus hypoadmissible approach regions (see [**KRA1**] or [**CIK**] for details of this terminology) are necessary devices in the study of lower-dimensional phenomena. We note (and this theme will recur in Section 2.2) that the role of hypoadmissible approach is not completely understood at this time. The metric approach that has been outlined here seems to shed some light on this matter and to suggest avenues for future work. ■

It is our view that the geometric framework provided by the Kobayashi/ Royden metric, or perhaps by the Sibony metric ([**SIB2**]) or by the convexification of one of these metrics, is the natural setting for function theory in several complex variables. In the one complex variable situation, this observation contributes nothing new (since in that context all domains are essentially the "same"). In the several variable setting, the use of an invariant metric picks out the natural non-isotropy of the situation and provides the necessary *analysis situs.*

1.7. Rigidity of Holomorphic Mappings. Let $\Omega \subset\subset \mathbb{C}^n$ be a bounded domain. Let $\Phi : \Omega \to \Omega$ be a holomorphic mapping. Let us consider the iterates $\Phi^j = \Phi \circ \Phi \circ \cdots \circ \Phi$. The study of the limit of the sequence $\{\Phi^j\}$, of the set of points on which it converges, and of the set on which it diverges, is sometimes termed "complex analytic dynamics." The discussion in this and the next section will be in the spirit of that subject, but the questions and techniques will be a bit different.

Let us begin by considering a point $P \in \Omega$ that is fixed by Φ.

PROPOSITION 1.7.1 (CARTAN). *If $\Phi'(P) = Jac_{\mathbb{C}}\Phi(P) = id$ then $\Phi(z) \equiv z$.*

PROOF. Since the techniques of this proof are influential throughout the discussion, we briefly discuss the proof.

Assume for simplicity that $P = 0$. If Φ is not the identity then the Taylor expansion of Φ about 0 will include terms other than the linear term. Thus

$$\Phi(z) = z + p_m(z) + \cdots ,$$

where $p_m(z)$ is a non-vanishing homogeneous polynomial of least order $m \geq 2$ in the expansion. [Note that this expansion is an expansion of n-tuples.] By direct calculation, we notice that

$$\Phi^j(z) = z + jp_m(z) + \cdots . \tag{1}$$

On the other hand, $\{\Phi^j\}$ is a normal family (since Ω is bounded) so there is a normally convergent subsequence $\{\Phi^{j_k}\}$. By Cauchy estimates it follows that the

m^{th} derivatives at the origin of the $\{\Phi^{j_k}\}$ converge. That contradicts (1); for the coefficients of jp_m *are* those m^{th} derivatives, and they blow up with j. We conclude that $\Phi(z) \equiv z$. ■

A companion result to Theorem 1.7.1 is

PROPOSITION 1.7.2. *If $\Phi : \Omega \to \Omega$, $\Phi(P) = P$, then none of the eigenvalues of $\Phi'(P)$ exceeds 1. In particular, the determinant of $\Phi'(P)$ has modulus not exceeding 1. If the modulus of the determinant equals 1, then Φ must be a biholomorphic mapping of Ω.*

PROOF. Assume that $P = 0$. Conjugate Φ with a change of coordinates so that $\Phi'(0)$ is in Jordan canonical form. If any of the eigenvalues (sitting above the diagonal) has modulus exceeding one, then iteration of Φ leads to a contradiction as in Cartan's theorem.

If all the eigenvalues have modulus equal to one, one finds (by diagonalization) a subsequence Φ^{j_k} such that $(\Phi^{j_k})'(0) \to \text{id}$. Then one can argue that $\Phi^{j_k} \to \text{id}$ normally. Finally, clever juggling of indices produces $\Phi^{j_m} \to f, \Phi^{k_\ell} \to g$ such that $f \circ g = \text{id}$ and $g \circ f = \text{id}$. From these identities, one can argue by contradiction that Φ is an automorphism. See [**KRA1**] for details. ■

Bedford [**BED1**] focused more sharply on this situation by considering iterates of an *automorphism* of Ω. He imposed the additional hypothesis that Ω be a domain of holomorphy and smooth, pseudoconvex (actually, taut will do). His principal result is

THEOREM 1.7.3. *Let $\Phi : \Omega \to \Omega$ be a biholomorphism of Ω. Let $\Phi^j \equiv \Phi \circ \Phi \circ \cdots \circ \Phi$ (j times) be as usual. Then precisely one of the following holds:*

- **1:** *$\{\Phi^j\}$ converges to a mapping that is the composition of an automorphism of Ω with retraction onto a (possibly lower dimensional) subvariety $V \subseteq \Omega$.*
- **2:** *$\{\Phi^j\}$ converges to a mapping of Ω into $\partial\Omega$.*

When one is studying an automorphism Φ of a domain belonging to a reasonable class, then it is not hard to determine when Φ actually has a fixed point. First we introduce some terminology.

DEFINITION 1.7.4. Let Ω be any domain. We define $\text{Aut}\,(\Omega)$ to be the group (under composition) of biholomorphic self-maps of Ω.

We topologize $\text{Aut}(\Omega)$ with the compact-open topology; this is the topology of uniform convergence on compact sets. Since normal family arguments are a fundamental tool in the subject, this is the natural topology to use.

In case Ω is bounded then elementary considerations show that $\text{Aut}\,(\Omega)$ is a (finite dimensional) real Lie group. In fact, fix $P \in \Omega$. Then the mapping

$$\text{Aut}\,(\Omega) \ni \phi \mapsto (\phi(P), \phi'(P))$$

parameterizes $\mathrm{Aut}(\Omega)$. (Here we are using the fact that any automorphism of Ω is an isometry of the Bergman metric, and any isometry is uniquely determined by its first order behavior.) This mapping shows simultaneously that $\mathrm{Aut}(\Omega)$ is finite dimensional*and* that it is a Lie group—without resort to the solution of Hilbert's fifth problem. A more general approach (using the Hilbert problem) may be found in [**KOB2**], where it is observed that any locally compact group of diffeomorphisms of a Euclidean domain is in fact a Lie group.

When Ω is bounded then $\mathrm{Aut}(\Omega)$ can never contain any complex structure. For if it did then it would give rise to a map of a complex disc in the Lie algebra into Ω (just fix $P \in \Omega$ and send x in the Lie algebra to $\exp_P x$). The group action could then be used to propagate this to a map of an entire complex line into Ω. That is impossible, by Liouville's theorem for instance (or by hyperbolicity if one wants to be more geometric about it). When Ω is unbounded then $\mathrm{Aut}(\Omega)$ is in general not a Lie group—for instance, $\mathrm{Aut}(\mathbb{C}^n)$ is infinite dimensional. Specifically, $\mathrm{Aut}(\mathbb{C}^2)$ contains the mappings

$$(z, w) \mapsto \big(z, w + f(z)\big)$$

for any holomorphic function f of one complex variable. Thus $\mathrm{Aut}(\mathbb{C}^2)$ is infinite dimensional. Slightly more interesting is the domain $\Omega = \{(z, w) \in \mathbb{C}^2 : \mathrm{Re}\, w > 0\}$. Then $\mathrm{Aut}(\Omega)$ contains the mappings

$$(z, w) \mapsto (z + f(w), \alpha w)$$

for any $\alpha > 0$, any holomorphic function f of the complex variable w. Again, $\mathrm{Aut}(\Omega)$ is infinite dimensional. See [**GK1**] for more on these matters. We see, for instance, that the infinite dimensionality of $\mathrm{Aut}(\Omega)$ in this second example makes it clear that Ω is not biholomorphic to a bounded domain. [Another way to see this is with the notion of hyperbolicity.]

DEFINITION 1.7.5. Let $\Phi \in \mathrm{Aut}(\Omega)$. We say that Φ is elliptic if the closed subgroup of $\mathrm{Aut}(\Omega)$ generated by Φ is compact.

In the category of strongly pseudoconvex domains, restricting attention to elliptic automorphisms is, in effect, a way to examine domains *other than the ball.* For the only strongly pseudoconvex domain with non-compact automorphism group is the ball. For weakly pseudoconvex domains, the question of which domains possess non-elliptic automorphisms is still open. A summary of some recent progress appears in [**KRA5**]. See also [**BEP1**]-[**BEP3**].

Important information about the topological properties of fixed point sets appears in [**VIG1**]-[**VIG3**]. Vigue gives sufficient conditions for the fixed point set to be connected, and other conditions that guarantee that all connected components of the fixed point set have the same dimension. Even in one complex dimension the fixed point sets of holomorphic self-maps of a domain can have some surprising properties. Here is an example:

THEOREM 1.7.6. *Let $\Omega \subseteq \mathbb{C}$ be a bounded domain. Let $\phi : \Omega \to \Omega$ be a holomorphic mapping (function). If ϕ has three distinct fixed points then ϕ is the identity mapping.*

The standard proof of this fact uses the uniformization theorem (see [**FAK**]). An alternative proof, by the present author, uses an analysis of Bergman metric geodesics and separation properties of planar geometry (the Jordan curve theorem, for instance). We instead concentrate here on the *existence* of fixed points.

One of the first results about fixed points, for a general class of domains, was the following:

THEOREM 1.7.7 (GREENE/KRANTZ [**GK2**]). *Let $\Omega = \{z \in \mathbb{C}^n : \rho(z) < 0\} \subseteq \mathbb{C}^n$ be strongly pseudoconvex. Let $\rho_0(z) = |z|^2 - 1$ be the defining function for the ball. If $k \in \mathbb{Z}^+$ is sufficiently large, $\epsilon > 0$ is sufficiently small, and $\|\rho - \rho_0\|_{C^k} < \epsilon$ then either Ω is biholomorphic to the ball or $\mathrm{Aut}\,(\Omega)$ has a fixed point $P \in \Omega$.*

One proves this result by checking that, under the hypotheses, the Bergman metric for Ω has strictly negative curvature. Next, if Ω is not the ball, then Bun Wong's theorem (see [**GK1**] for an exposition) implies that Ω has compact automorphism group. Finally, one invokes a classical fixed point result of Hadamard (the fixed point is identified as a center of gravity for the metric).

I suspect that the theorem, and more, is true in the C^2 topology but I do not know how to prove it using the methods of [**GK2**]. However this issue is obviated by the following much stronger result:

THEOREM 1.7.8 (LEMPERT [**LEM2**]). *Let Ω be any bounded, convex domain and G a compact group of automorphisms. Then G has a fixed point.*

PROOF. Instead of the Bergman metric, we use the Kobayashi metric. By Lempert's important work [**LEM1**], metric balls are then known to be convex. Let k denote the integrated Kobayashi distance on Ω. For $z \in \Omega$ and $r > 0$ we let

$$\bar{B}(z, r) \equiv \{w \in \Omega : k(z, w) \leq r\}.$$

If $z \in \Omega$ then let Gz denote its orbit under the action of G. Observe that if $R > 0$ is large enough then $K_R(z) \equiv \bigcap\{\bar{B}(w, R) : w \in Gz\}$ is a compact, convex, non-void and G-invariant set in Ω. We set

$$r = \inf\left\{R > 0 : K_R(z) \neq \emptyset\right\}.$$

Then $K = K_r(z)$ has all these properties as well. It is also true, by the minimality of r, that the interior of $K_r(z)$ is empty. Thus $\dim K_r(z) < 2n$.

If $K_r(z)$ contains just one point then we are done—this one point is the fixed point we desire. Therefore we must address the possibility that $K_r(z)$ contains more than one point.

Let z' be any element of $K_r(z)$ other than z itself. Set

$$r' = \inf\{R : K_r(z) \cap K_R(z') \neq \phi\}.$$

Consider the set $K' = K \cap K_{r'}(z')$. Then K' is compact, convex, non-empty, and G-invariant. It is also the case that $\dim K' < \dim K$. This last assertion follows from the fact that the affine hull H of K intersects each $\partial B(w,r)$ in a $(\dim K)-1$ dimensional surface. Thus if K' had an interior point relative to H then this point would be at a distance less than r' from Gz'. This would contradict the minimality of r'.

We thus obtain a sequence $K \supseteq K' \supseteq K'' \cdots$ of compact, convex, non-empty, G-invariant sets of strictly decreasing dimension. After at most $2n$ steps we obtain such a set that has dimension 0. By convexity, this set must be a singleton. Its point is the fixed point that we seek. ∎

Many closely related results about semi-continuity of automorphism groups fail when the boundary fails to be C^1 smooth. The next example shows that Theorem 1.7.7, strictly interpreted, also fails in this setting:

EXAMPLE 1.7.9. Let $B = \{z \in \mathbb{C}^2 : \rho(z) = |z_1|^2 + |z_2|^2 - 1 < 0\}$ be the unit ball. Let $\phi \in C_c^\infty(\mathbb{C}^2)$ be supported in $B(0, 1/50)$ and positive on $B(0, 1/100)$. Set $\tilde{\rho}(z) = \rho(z) - \phi(z_1 - i, z_2)$. Then

$$\tilde{\Omega} = \{(z_1, z_2) : \tilde{\rho}(z) < 0\}$$

is a smoothly bounded domain. Set

$$\Phi(z_1, z_2) = \left(\frac{z_1 + 1/3}{1 + (1/3)z_1}, \frac{\sqrt{8/9}z_2}{1 + (1/3)z_1}\right).$$

Define

$$\Omega^* = \bigcup_{j=1}^{\infty} \Phi^j(\tilde{\Omega}).$$

Then Ω^* is the ball with a sequence of bumps receding to the point $(1,0)$. One may check that, if $\|\phi\|_{C^1}$ is small then the domain Ω^* is $C^{1-\epsilon}$ close to the ball. See Figure 1.15.

One may also check that Ω^* is *not biholomorphic* to the ball. To see this, suppose that it were. Let Ψ denote the biholomorphism. Then it is easy to argue that a portion of the boundary of Ω^* that is isometric to the ball would be mapped under Ψ to a piece of the boundary of the ball. By Alexander's theorem [**ALE**], it would follow that Ψ is a (globally) linear fractional automorphism of the ball. But then, in particular, Ψ could not be a biholomorphism of Ω^* to the ball.

Finally, notice that Aut (Ω^*) has non-compact automorphism group, because it contains the elements Φ^j and these converge to a constant mapping (which is not in the automorphism group). In particular, it does not have a fixed point. So it fails the conclusion of the last theorem. ∎

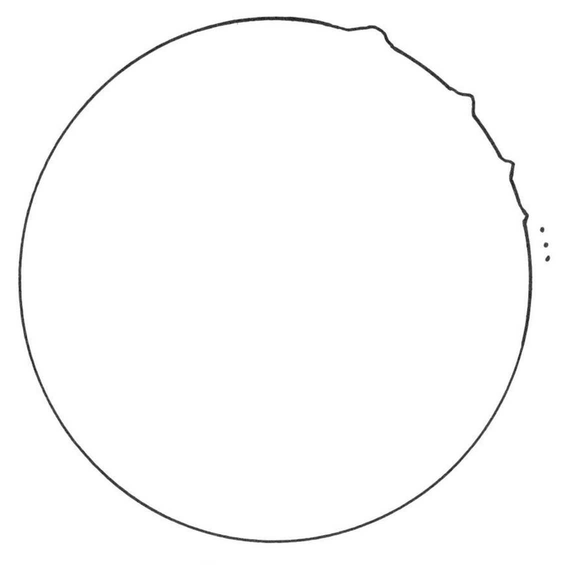

FIGURE 1.15

Next we turn to questions about fixed points of a single mapping. The classical result is that of Denjoy-Wolff:

THEOREM 1.7.10. *Let $f : D \to D$ be a self-mapping (not necessarily an automorphism) of the unit disc. Assume that f is not the identity. Then the sequence of iterates f^k does not converge if and only if f is an automorphism of D with exactly one fixed point. Furthermore, when the limit of f^k exists then that limit is a constant function with absolute value not exceeding 1.*

M. Abate has succeeded in generalizing this result to several types of domains in higher dimensions. Complete references may be found in [**ABA3**]. See also [**ABA1**],[**ABA2**].

THEOREM 1.7.11 (ABATE). *Let $\Omega \subseteq \mathbb{C}^n$ be a convex domain with C^2 boundary. Assume that $f : \Omega \to \Omega$ is holomorphic and has no fixed points. Fix $P \in \Omega$ and suppose that $Q \in \bar{\Omega}$ is a limit point of the sequence $\{f^k(P)\}$. Then $Q \in \partial\Omega$.*

Moreover, if h is any limit of a subsequence of the functions $\{f^k\}$ then $h(\Omega) \subseteq H_Q$, where H_Q is the intersection of $\partial\Omega$ with the supporting hyperplane to Ω at Q. In particular, if Ω is strictly convex then $\{f^k\}$ converges to a single point in $\partial\Omega$.

THEOREM 1.7.12 (ABATE). *Let $\Omega \subseteq \mathbb{C}^n$ be taut. Let $f : \Omega \to \Omega$ be a holomorphic mapping with a fixed point $P \in \Omega$. Then the sequence of iterates $\{f^k\}$ converges if and only if the Jacobian matrix $f'(P)$ has no eigenvalue λ such that $|\lambda| = 1$ and $\lambda \neq 1$.*

In his proofs, Abate makes decisive use of (i) the holomorphic contractibility of a convex domain and (ii) the fact that Kobayashi metric balls in a convex domain are convex. Both of these properties fail for a general strictly pseudoconvex domain—even one that is contractible. And we might expect to make topogical restrictions on our domains if we hope to find a fixed point (in view of the Lefschetz fixed point theorem, for example).

Daowei Ma [**MA1**] found a way around these limitations using delicate calculations of the Kobayashi metric. Two of his theorems are as follows:

THEOREM 1.7.13 (MA). *Let $\Omega \subseteq \mathbb{C}^2$ be a contractible, strongly pseudoconvex domain with smooth boundary, and let $f : \Omega \to \Omega$ be a holomorphic self-map without fixed points. Then the sequence of iterates $\{f^k\}$ converges to a point on in $\partial\Omega$.*

THEOREM 1.7.14 (MA). *Let $\Omega \subseteq \mathbb{C}^n$ be a contractible, strongly pseudoconvex domain with smooth boundary. Assume that Ω is not biholomorphic to the ball. Then every automorphism of Ω has a fixed point in Ω.*

The proofs of these results are very technical, but the underlying geometry is pretty. We sketch the ideas.

SKETCH OF PROOF OF THEOREM 1.7.14: Since Ω is not biholomorphic to the ball, its automorphism group G must be compact (this fact, due to Bun Wong, will be discussed in detail in Section 1.10). If m is bi-invariant Haar measure on G, then we may average the distance function to $\partial\Omega$ (which is smooth near $\partial\Omega$) to obtain a function h near $\partial\Omega$ such that $h\big|_{\partial\Omega} = 0$ and so that the level sets of h are invariant under the action of G. But then the sublevel set Ω_ϵ of h corresponding to $h = \epsilon$, $\epsilon > 0$ small, will be invariant under the action of G. By Morse theory, one can see that Ω_ϵ will still be contractible if ϵ is small enough. If f is an automorphism of Ω then the Brouwer fixed point theorem applies to f acting on Ω_ϵ. We find a fixed point $P \in \Omega_\epsilon \subset\subset \Omega$. The proof is complete. ■

SKETCH OF PROOF OF THEOREM 1.7.13: The hard technical part of the proof is to use estimates of horospheres to see that any boundary accumulation point of the iterates must be unique. This involves delicate estimates using an approximate holomorphic retraction. We shall not treat the details.

Taking the result of the first paragraph of the proof for granted, we see that if the theorem were false then a subsequence of the $\{f^k\}$ converges to a function $h : \Omega \to \Omega$. Using the Theorem 1.7.3 of Bedford, we conclude that $h = g \circ \rho$, where $g \in \mathrm{Aut}\,(\Omega)$ and ρ is a holomorphic retraction onto a subvariety X of Ω. Moreover, ρ is the limit of a subsequence of the $\{f^k\}$. Since $\rho \circ f = f \circ \rho$, we see that $f(X) \subseteq X$. Set $f_1 = f\big|_X$. Then a subsequence of $\{f_1^k\}$ converges to id_X. So $f_1 \in \mathrm{Aut}\,(X)$. Now we have three cases:

(i) $\dim(X) = 0$. Then X consists of a single point. This point must be a fixed point of f, and we have a contradiction.

(ii) $\dim(X) = 1$. Then X is a contractible, complete hyperbolic Riemann surface. Thus, by uniformization, X is (conformally equivalent to) the unit disc. Since a subsequence of $\{f_1^k\}$ converges to the identity then the Denjoy-Wolff theorem implies that f_1 has a fixed point. But then f has a fixed point in X. That is a contradiction.
(iii) $\dim(X) = 2$. Then $X = \Omega$ and $f \in \mathrm{Aut}(\Omega)$. Now a subsequence of $\{f^k\}$ converges to $\rho = \mathrm{id}$. In case Ω is biholomorphic to the ball we may take Ω to be convex and may then conclude from Abate's first theorem that f has a fixed point: contradiction. If Ω is not biholomorphic to the ball then we may apply Theorem 1.7.14 to see that f has a fixed point: again a contradiction.

Having covered all cases, we see that the theorem is proved. ∎

I do not know at this time whether an inductive argument can be constructed to prove the first theorem in all dimensions.

We would like to conclude by mentioning some more recent ideas of X. Huang. His ideas capitalize on some recent developments connected with the metric developed by Kobayashi/Royden and, as such, fit into the mainstream of the ideas in this book.

THEOREM 1.7.15 (X. HUANG [**HUA1**]). *If Ω satisfies Bell's Condition R, if $\bar{\Omega}$ is diffeomorphic to the ball, and if $\Phi \in \mathit{Aut}(\Omega)$ is elliptic, then Φ has an interior fixed point.*

The proof of this theorem combines ideas from the preceding two results in a clever way.

COROLLARY 1.7.16. *Let $\Omega \subset\subset \mathbb{C}^n$ be smoothly bounded, pseudoconvex, and have finite type boundary. If $\sigma \in \mathit{Aut}(\Omega)$ fixes three points of $\partial\Omega$ then σ has an interior fixed point.*

This result is remarkable in that the hypothesis is independent of dimension (Remember that the determination of three boundary points for an automorphism of the disc uniquely determines the automorphism. Also, in dimension one, three interior fixed points makes the mapping the identity. These statements are false in higher dimensions.). The proof entails showing that the hypothesis implies that σ is elliptic. Naturally, Denjoy-Wolff theory is involved.

1.8. Rigidity at the Boundary. Now we treat ideas connected to fixed points in the boundary. This topic is inspired, at least philosophically, by Fatou-Julia-Denjoy-Wolff theory.

The boundary rigidity phenomenon that we treat lies at once nearer to the heart of the function theory of several complex variables and also nearer to the surface (because it is more elementary in nature). However we believe that it is linked to deeper phenomena that are not yet fully understood.

Let $\Omega \subset\subset \mathbb{C}^n$. Here the case $n = 1$ is already of some interest. Fix a point $P \in \partial\Omega$ and a small neighborhood U of P in $\mathbb{C}^n$. Let $\Phi : \Omega \to \Omega$ and assume that Φ continues to $\partial\Omega \cap U$ in some sense (so that we may speak of $\Phi(U \cap \partial\Omega)$).

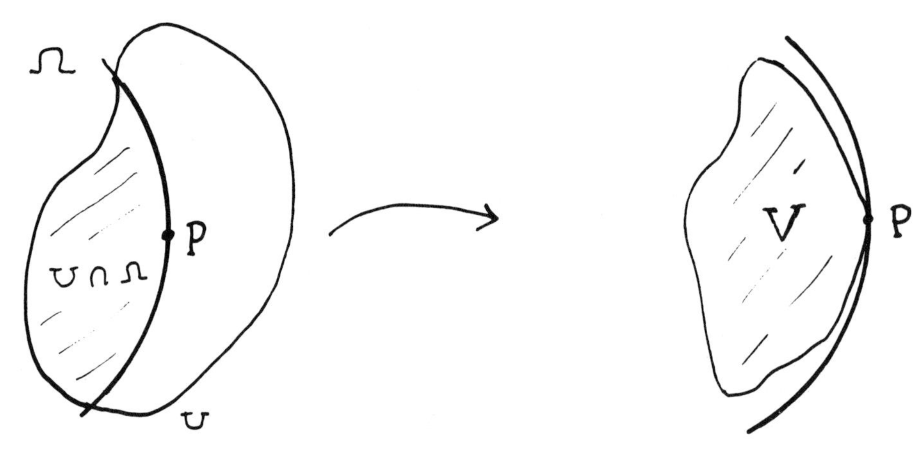

FIGURE 1.16

Assume that $\Phi(P) = P$. Is it possible for $\Phi(U \cap \partial\Omega) \equiv V$ to have high order of contact with $\partial\Omega$ at P? Refer to Figure 1.16.

This question was originally put to me by Warren Wogen, who had tantalizing applications to the theory of composition operators in mind. I was hampered in its solution (it was finally solved in the joint work [**BUK**] with D. Burns) by the imprecision of the formulation of the problem. If the question is interpreted naively, then one sees that in one complex dimension with $\Omega = D$ (the disc) the answer is that arbitrarily high order of contact is possible. Indeed (see Figure 1.17), the domain V may be smoothly bounded and have arbitrarily high (even infinite) order of osculation with $\partial\Omega$ at P. Because V is simply connected, we may map Ω conformally onto V (this conformal mapping is the map Φ that we seek) and thus achieve any desired "order of contact."

However there is no Riemann mapping theorem in several complex variables (see the discussion of this matter in [**GK1**]). And certain other well-known results about rigidity of holomorphic mappings in several complex variables suggest that matters may be different in that context. For motivation consider the result of H. Alexander [**ALE**]:

THEOREM 1.8.1. *If $U \cap B$ and $U' \cap B$ are boundary neighborhoods in the ball and if Φ is a biholomorphic mapping of these neighborhoods which extends C^2 to the boundary, then Φ must be the restriction to $U \cap B$ of a biholomorphism of the entire ball.*

S. Pinchuk [**PIN**] generalized this result to bounded strictly pseudoconvex domains with real analytic boundaries and W. Rudin [**RUD1**] reduced the hypothesis of C^2 to the boundary to an assumption which is even weaker than continuity at a point. There are some generalizations of a different nature that appear in [**GK6**]. We briefly indicate these latter results:

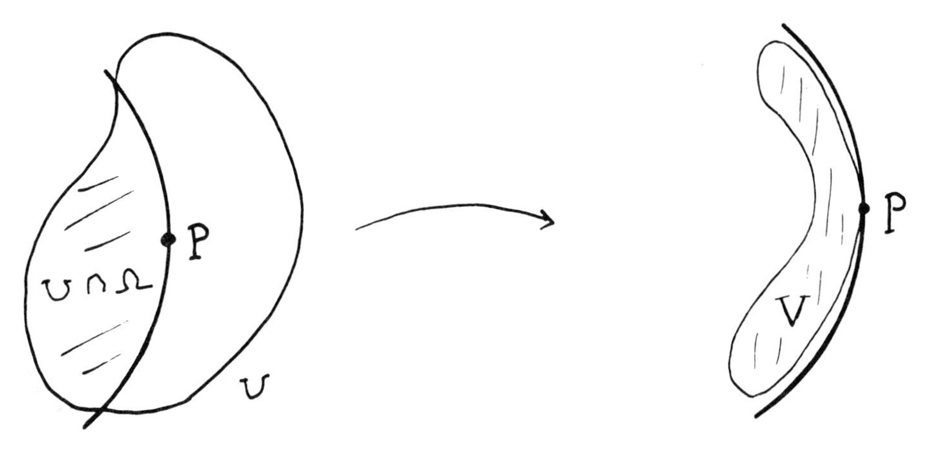

FIGURE 1.17

PROPOSITION 1.8.2. *Let Ω be a strongly pseudoconvex domain with real analytic boundary. Let $P \in \partial\Omega$ and let U be a small neighborhood of P in the ambient space.*

Assume that

$$f : U \cap \bar{\Omega} \to U \cap \bar{\Omega}$$

is a C^2 diffeomorphism with $f(U \cap \partial\Omega) \subseteq U \cap \partial\Omega$. Further assume that

$$\|f\|_{C^{2+\epsilon}} \leq M_0,$$

for some constant $M_0 > 0$. Let $K \subset\subset U \cap \Omega$ and $L \subset\subset U \cap \Omega$. Assume that $f(K) \subseteq L$. If $\bar{\partial} f$ is C^0 sufficiently small (depending on M_0, K, L, ϵ) then f is C^2 close to being the restriction to $U \cap \Omega$ of an automorphism of Ω.

PROPOSITION 1.8.3. *Let Ω be strongly pseudoconvex, bounded, and smooth. Let Φ be a diffeomorphism of the closure of Ω into (but not necessarily onto) the closure of the ball. Assume that $\|\bar{\partial}\Phi\|_{C^{1+\epsilon}}$ is small. Further suppose that $P_0 \in \partial\Omega$ and that there is a boundary neighborhood $U \ni P_0$ such that $U \cap \partial\Omega$ is $C^{2+\epsilon}$ close to ∂B. Then $\partial\Omega$ is globally uniformly close to ∂B.*

Taken together, 1.8.1-1.8.3 suggest (albeit somewhat tenuously) that infinite order of contact in our original mapping problem is out of the question in the several complex variable setting. It is also the case that the natural dilation structure of the Heisenberg group, which is canonically equivalent to the boundary of the ball (see [**BCK**]) is inconsistent with order of contact higher than four.

It turns out that the most efficient way to examine the questions being raised here is to change them. In the present context, the geometric approach (while attractive and suggestive) is too vague. Thus we formulate a positive result in terms of *analytic order of contact.* We close this section with three statements about analytic uniqueness theorems at the boundaries of domains in one and several complex variables. We shall briefly discuss their proofs. In the next section we shall discuss the geometric significance of these results, and consider some generalizations. Finally, we return to and answer the original *geometric* question.

THEOREM 1.8.4 (BURNS/KRANTZ [**BUK**]). *Let $\Phi : D \to D$ be a holomorphic mapping of the disc to itself such that*

$$\Phi(\zeta) = \mathbb{1} + (\zeta - \mathbb{1}) + \mathcal{O}\left(|\zeta - \mathbb{1}|^4\right)$$

as $\zeta \to 1$. Then $\Phi(z) \equiv z$ on the disc.

Notice that this result is in the vein of the uniqueness part of the Schwarz lemma, formulated at the boundary point 1. The hypothesis in the displayed equation is to the effect that Φ agrees with the identity at 1 to fourth order. It turns out that this order of agreement cannot be meaningfully reduced without the imposition of further hypotheses. An example to show that the result is sharp will be discussed below.

We mention that, had we known about it, our Theorem 1 might have been inspired by [**VEL**]. In that source Velling proves a similar result with the additional crucial hypothesis that ϕ is analytic in a neighborhood of 1. This assumption enables him to use certain deformation ideas that go back to Löwner. We sketch his proof at the end of the section. Also, Sibony has pointed out that the exponent 4 is rather natural when the theorem is analyzed from the point of view of the "Julia rose." We discuss this remark at the end of the section as well.

THEOREM 1.8.5 (BURNS/KRANTZ [**BUK**]). *Let $\Phi : B \to B$ be a holomorphic mapping of the ball to itself such that*

$$\Phi(z) = \mathbb{1} + (z - \mathbb{1}) + \mathcal{O}\left(|z - \mathbb{1}|^4\right)$$

as $z \to 1$. (Here $\mathbb{1}$ denotes the distinguished boundary point $\mathbb{1} = (1, 0, \ldots, 0)$ of the ball .) Then $\phi(z) \equiv z$ on the ball.

THEOREM 1.8.6 (BURNS/KRANTZ [**BUK**]). *Let $\Omega \subseteq \mathbb{C}^n$ be a strongly pseudoconvex domain with smooth boundary. Fix $P \in \Omega$. Let $\Phi : \Omega \to \Omega$ be a holomorphic mapping of Ω to itself such that*

$$\Phi(z) = P + (z - P) + \mathcal{O}\left(|z - P|^4\right)$$

as $z \to P$. Then $\phi(z) = z$ on Ω.

The key analytic ideas are contained in the proof of Theorem 1.8.4. In fact it can be proved in several ways. At this time we shall discuss two of these.
PROOF OF THEOREM 1.8.4: Consider the holomorphic function

$$g(\zeta) = \frac{1 + \phi(\zeta)}{1 - \phi(\zeta)}.$$

Then g maps the disc D to the right half plane. By the Herglotz representation (this is just an application of the Banach-Alaoglu theorem, or see also [**AHL3**] for another proof), there must be a positive measure μ on the interval $[0, 2\pi)$ and an imaginary constant $\mathcal{C}$ such that

$$g(\zeta) = \frac{1}{2\pi}\int_0^{2\pi} \frac{e^{i\theta}+\zeta}{e^{i\theta}-\zeta}\,d\mu(\theta) + \mathcal{C}. \tag{1}$$

We use the hypothesis on ϕ to analyze the structure of g and hence that of μ. To wit,

$$\begin{aligned}
g(\zeta) &= \frac{1+\zeta+\mathcal{O}(\zeta-1)^4}{1-\zeta-\mathcal{O}(\zeta-1)^4} \\
&= \frac{1+\zeta+\mathcal{O}(\zeta-1)^4}{(1-\zeta)(1+\mathcal{O}(\zeta-1)^3)} \\
&= \frac{1+\zeta}{1-\zeta}\frac{1}{1+\mathcal{O}(1-\zeta)^3} + \mathcal{O}(1-\zeta)^3 \\
&= \frac{1+\zeta}{1-\zeta}\cdot\left(1+\mathcal{O}(1-\zeta)^3\right) + \mathcal{O}(1-\zeta)^3 \\
&= \frac{1+\zeta}{1-\zeta} + \mathcal{O}(\zeta-1)^2. \qquad (2)
\end{aligned}$$

From this and equation (1) we easily conclude that the measure μ has the form

$$\mu = \delta_0 + \nu, \tag{3}$$

where δ_0 is (2π times) the Dirac mass at the origin and ν is another positive measure on $[0, 2\pi)$. In fact a nice way to verify the positivity of ν is to use the equation

$$\frac{1+\zeta}{1-\zeta} + \mathcal{O}(\zeta-1)^2 = \frac{1}{2\pi}\int_0^{2\pi} \frac{e^{i\theta}+\zeta}{e^{i\theta}-\zeta}\,d(\delta_0+\nu)(\theta) + \mathcal{C} \tag{4}$$

to derive a Fourier-Stieltjes expansion of $\delta_0 + \nu$ and then to apply the Herglotz criterion ([**KAT**], p. 38).

We may use (3) to simplify equation (4) to

$$\mathcal{O}(\zeta-1)^2 = \frac{1}{2\pi}\int_0^{2\pi} \frac{e^{i\theta}+\zeta}{e^{i\theta}-\zeta}\,d\nu(\theta) + \mathcal{C}.$$

Pass to the real part of the last equation, thus eliminating the constant $\mathcal{C}$. Since ν is a positive measure, we thus see that real part of the integral on the right of this last equation represents a positive harmonic function h on the disc that satisfies

$$h(\zeta) = \mathcal{O}(\zeta-1)^2.$$

In particular, h takes a minimum at the point $\zeta = 1$ and its gradient vanishes there as well. This contradicts Hopf's lemma ([**KRA1**] or [**KRA7**]) unless $h \equiv 0$. But $h \equiv 0$ implies that $\nu \equiv 0$ hence that

$$g(\zeta) = \frac{1+\zeta}{1-\zeta}.$$

Therefore $\phi(\zeta) \equiv \zeta$, and the theorem is proved. ■

SECOND PROOF OF THEOREM 1.8.4: With ϕ as in Theorem 1.8.4, we have by equation (2) in the first proof that

$$\begin{aligned} g(\zeta) &= \frac{1+f(\zeta)}{1-f(\zeta)} \\ &= \frac{1+\zeta}{1-\zeta} + \mathcal{O}\left((\zeta-1)^2\right). \end{aligned}$$

Thus

$$\exp(-g(\zeta) = \exp\left(-\frac{1+\zeta}{1-\zeta}\right) \cdot \exp\left[\mathcal{O}(\zeta-1)^2\right].$$

As a result,

$$\exp\left[\mathcal{O}(\zeta-1)^2\right] = \exp(-g(\zeta)) \cdot \exp\left(\frac{1+\zeta}{1-\zeta}\right).$$

The second expression on the right has unimodular boundary values almost everywhere and the first is bounded in absolute value by 1 (since $\mathrm{Re}\, g > 0$). Thus the expression $H(\zeta) \equiv \exp(\mathcal{O}(\zeta-1)^2)$ is bounded in absolute value by 1, attains a maximum at $\zeta = 1$, and has vanishing gradient at that point. The Hopf Lemma is contradicted unless $\mathcal{O}(\zeta-1)^2 \equiv 0$ or $H(\zeta) \equiv 1$. Thus

$$g(\zeta) = \frac{1+\zeta}{1-\zeta}$$

and we conclude that $\phi(\zeta) = \zeta$. ■

The second proof shows, without reference to the Herglotz theorem, that the function ϕ can be suitably normalized, and have its "principal part" subtracted off, so that a maximum principle (i.e. Hopf lemma) argument can be applied directly.

In fact the proofs just presented work if only

$$\Phi(z) = P + (z-P) + o\left(|z-P|^3\right),$$

However this is essentially the weakest possible hypothesis. For the function

$$\phi(\zeta) = \zeta - \frac{1}{10}(\zeta-1)^3.$$

maps the disc to the disc (as one may check with a calculation), satisfies the hypothesis of the theorem with exponent 3, yet is plainly not the identity.

PROOF OF THEOREM 1.8.5: There is no useful Herglotz representation on the ball (however, see [**AIZ2**] for related ideas): this is a deep fact which cannot be circumvented. Thus we present a new argument that reduces the ball case to the disc case. For simplicity we restrict attention to dimension two.

For each point $a \in B$ let ℓ_a be the complex linear span of a and $\mathbb{1}$. Let d_a be the complex disc given by $\ell_a \cap B$. Now for fixed a consider the holomorphic function

$$\begin{aligned} \psi : D &\rightarrow B \\ \zeta &\mapsto (\zeta, 0). \end{aligned}$$

Define also the mapping

$$\phi_a : B \rightarrow B,$$

which is an automorphism of the ball mapping d_0 onto d_a. (That such maps exist follow from elementary geometric considerations: just use a Möbius transformation to adjust the distance to the boundary and a rotation to modify the inclination; alternatively, see [**RUD1**] for an explicit formula.). Finally define

$$\begin{aligned} \pi_1 : B &\rightarrow B \\ (z_1, z_2) &\mapsto (z_1, 0); \end{aligned}$$

$$\begin{aligned} \eta : d_0 &\rightarrow D \\ (z_1, 0) &\mapsto z_1. \end{aligned}$$

Then the function

$$\begin{aligned} H_a : D &\rightarrow D \\ \zeta &\mapsto \eta \circ \pi_1 \circ (\phi_a)^{-1} \circ \Phi \circ \phi_a \circ \psi(\zeta) \end{aligned}$$

is well-defined. Moreover, it is straightforward to check that H_a satisfies the hypotheses of Theorem 1.8.4 (since Φ agrees with the identity to high order, the composition $(\phi_a)^{-1} \circ \Phi \circ \phi_a$ is the identity to high order). It follows that $H_a(\zeta) \equiv \zeta$.

Now set

$$\begin{aligned} G_a(\zeta) &= (\phi_a)^{-1} \circ \Phi \circ \phi_a \circ \psi(\zeta) \\ &\equiv (g_a^1(\zeta), g_a^2(\zeta)). \end{aligned}$$

The statement that $H_a(\zeta) \equiv \zeta$ means that $g_a^1(\zeta) \equiv \zeta$. But

$$|g_a^1(\zeta)|^2 + |g_a^2(\zeta)|^2 < 1$$

for $\zeta \in D$. Letting $|\zeta| \rightarrow 1$ yields then that $|g_a^2(\zeta)| \rightarrow 0$. Thus $g_a^2 \equiv 0$. It now follows that the image of G_a already lies in d_0. As a result, it must be that Φ preserves d_a. But this can only hold for every choice of a if Φ is the identity mapping.

The proof is now complete. ∎

We note that the second proof of Theorem 1 can be adapted without difficulty to the ball.

For Theorem 1.8.6, the essential observation is that we really did not use the structure of the ball in any essential way in the proof of Theorem 1.8.5. In fact all that was really needed was a way to map discs into the domain, and a way to map back out again, such that the composition of the two mappings was the identity. If Ω is strongly convex, then Lempert's theory of extremal discs for the Kobayashi metric [**LEM3**] provides the necessary machinery. Essential in this approach is his results (i) that extremal discs are regular to the boundary and (ii) that the domain may be holomorphically *retracted* onto each extremal disc (see [**LEM3**].

If Ω is strongly *pseudo*convex, then it is well known that there is no analogue for Lempert's discs (see [**SIB1**]). However, as noted on pages 467-8 of [**LEM1**], the extremal discs for near tangent directions at a boundary point are in fact local: if $P \in \partial\Omega$, if U is a small neighborhood of P, and if ξ points into the domain and is nearly tangent to $\partial\Omega$ at P, then the Kobayashi extremal disc for Ω at P in the direction ξ is just the same as the Kobayashi extremal disc for $\Omega \cap U$ at P in the direction ξ. This turns out to be enough to push through the preceding proof for strongly pseudoconvex domains.

We repeat that it is crucial in these theorems to keep track of the local/global dialectic. If Ω is the ball in $\mathbb{C}^n$, Φ is a univalent holomorphic mapping on $\bar{\Omega}$ with smooth image, and Ω' is a smooth domain that contains $\Phi(\Omega)$ and that osculates $\Phi(\Omega)$ to N^{th} order at $\Phi((1,0,\ldots,0))$ $(N >> 6)$ then we certainly cannot conclude that Φ is the identity. In fact, in this construction Φ is perfectly arbitrary. And of course (and this is the point) $\Omega' \neq \Omega$.

In general, even if the domain and target of the mapping Φ are the same, one cannot allow the boundary point P to be mapped to a *different* boundary point P'. However, imagine the situation in which we know *a priori* that there is an automorphism μ of our strongly pseudoconvex domain that takes the boundary point P to the boundary point P'. Further suppose that we are considering a holomorphic mapping $\Phi : \Omega \to \Omega$ that takes P to P' in some reasonable sense and that the image osculates the target at Q to high order. Then we may apply theorem 1.8.6 above to the function $\Phi \circ \mu^{-1}$ to conclude that $\Phi = \mu$.

In fact the result just stated can be forced to be true *without* the assumption that μ exists. If one assumes that the domain in question has real analytic boundary, and that the boundary point P is mapped by Φ to the boundary point P' with high order of osculation, then a delicate analytic continuation argument shows that Φ is an automorphism of Ω. Details are may be found in [**BUK**].

We close this section with a discussion of other ways to prove Theorem 1.8.4 and some other points of view. In the next section we will return to Theorems 1.8.5 and 1.8.6.

Turn again to the setting of the disc in dimension one. As previously noted, one way to think of this one variable result is by way of the so-called Julia rose. Suppose that the mapping coordinates have been normalized so that the domain point $1 + i0$ is sent to the origin. Suppose further that our domain, the disc, has been mapped to the upper half plane. Our point of view now is to assume that Φ is holomorphic *in a neighborhood of the origin*, that $\Phi(0) = 0$, and that $\Phi'(0) = 1$. Thus we are supposing that

$$\Phi(\zeta) = \zeta + \zeta^p + \cdots$$

for some integer $p > 1$. We wish to discover, from geometric considerations, what is the least exponent p that forces the conclusion of Theorem 1.8.4.

Let ξ be a primitive $(2p-2)^{th}$ root of unity: the equation $\xi^{2p-2} = 1$ implies that $\xi^{p-1} = -1$ hence $\xi^p = -\xi$. Then for any integer $0 \leq k < 2p-2$ we know that

$$(\xi^k)^p = (-1)^k \xi^k.$$

Now consider $\zeta = t\xi^k$ for t a small real parameter. Then

$$\begin{aligned} \Phi(\zeta) &= \Phi(t\xi^k) \\ &= t\xi^k + (t\xi^k)^p \\ &= t\xi^k + t^p(-1)^k\xi^k \\ &= (t + (-1)^k t^p)\xi^k. \end{aligned}$$

This calculation shows that, in the direction ξ^k, we have that

$$\Phi \sim \mathrm{id} + (-1)^k t^p.$$

Thus we see that, in the directions generated by ξ, f is alternatively attractive and repellent. We know from Fatou-Julia-Denjoy-Wolff theory that a boundary fixed point for a self map of the half plane (resp. the disc), such as the origin is for us, can be either attractive or repellent but not both. [Since our mapping Φ satisfies $\Phi'(0) = 1$ then the point 0 must in fact be attractive.] This assertion is consistent with the analysis we have just performed *provided that* $p \leq 3$. See Figure 1.18. However as as soon as $p \geq 4$ then $2p - 2 \geq 6$ and we see that two successive leaves of the rose must fall in the upper half plane—one of them attractive and one of them repellent. See Figure 1.19. That is a contradiction.

Yet another way to look at Theorem 1.8.4 comes from [**VEL**]. We again assume that Φ is analytic in a neighborhood of 1, but we work with the disc as given. We further assume, for convenience, that Φ is univalent. Let $\Omega \subseteq D$ be the image of the function Φ. The picture is given in Figure 1.20. Seeking a contradiction, we assume that Φ is not onto (if it were onto, then it would have to be a Möbius transformation and then, by the hypothesis on the asymptotic behavior at 1, would have to be the identity).

Notice that in Figure 1.21 we have inserted a radial slit S_1 from $\partial\Omega$ to the circle. Let g_1 be the conformal mapping uniformization of $D \setminus S_1$ onto the disc, normalized so that $g_1'(1) = 1$ and $g_1''(1) = 0$. Let Ω_1 be the image of Ω under g_1.

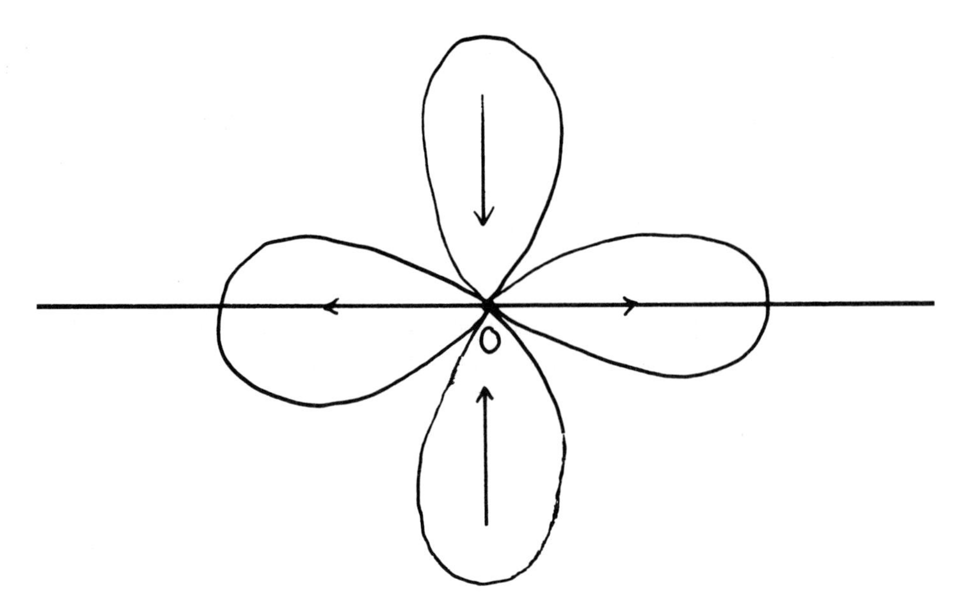

FIGURE 1.18

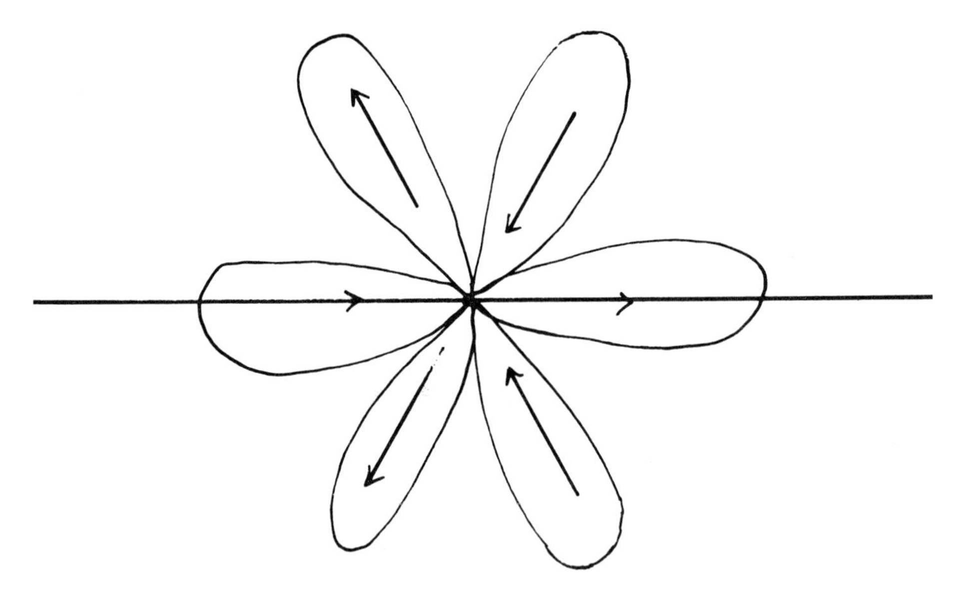

FIGURE 1.19

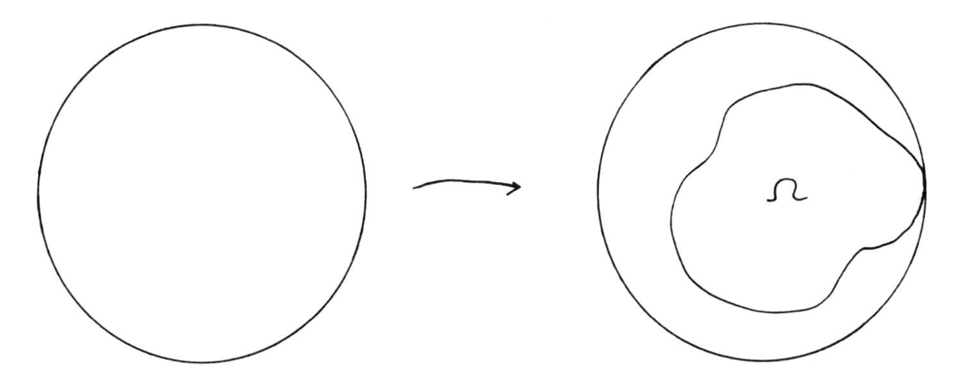

FIGURE 1.20

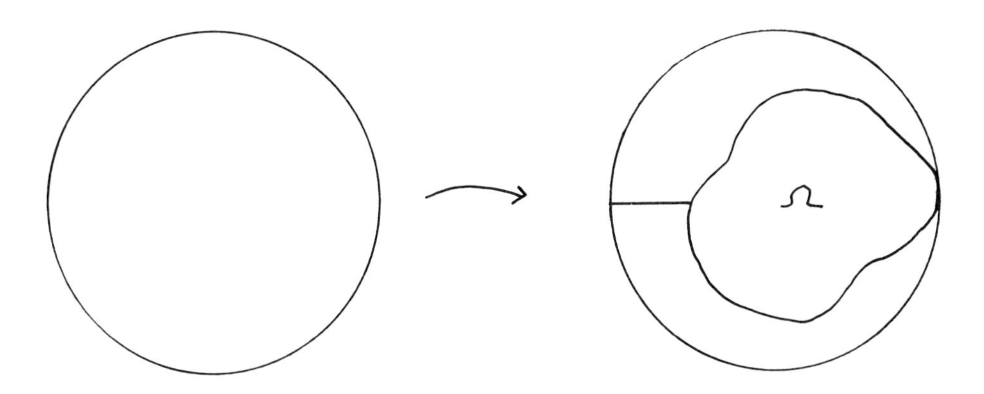

FIGURE 1.21

Now repeat this process with Ω_1 replacing Ω and a radial slit S_2 extending from $\partial\Omega_1$ out to the circle. Continue.

Using some old ideas of Löwner about distortions of mappings, it can be seen that the domains Ω_j exhaust the disc. Because of our normalizations, it can be calculated that

$$\begin{array}{rcl} g_1'''(1) & < & 0 \\ (g_2 \circ g_1)'''(1) & < & g_1'''(1) \\ & \text{etc.} & \end{array}$$

Putting these ideas together, we extract from the normal family $\mu_k \equiv g_k \circ g_{k-1} \circ \cdots \circ g_1$ a normally convergent sequence converging to a conformal mapping g of Ω to the disc and satisfying

$$g(\zeta) = \zeta + \mathcal{O}\big((\zeta - 1)^3\big),$$

with the coefficient of the cubic term being negative. But then the composition $h \equiv g \circ \Phi$ is a conformal mapping of the disc to the disc that satisfies

$$h = \zeta + \mathcal{O}\big((\zeta - 1)^3\big).$$

By our construction, the coefficient of the cubic term is negative. If we think of h as the composition of a rotation and a Möbius transformation, then we must conclude that the rotation is the identity. Then a simple calculation forces the Möbius transformation to be the identity as well. So h must be the identity. That contradicts our information about the third derivative. The result is established.

It is my view that there is a great deal of interesting function theory lurking in the background here. In fact another proof of Theorem 1.8.4 (the original one discovered by Burns) examines the boundary asymptotics of Pick's inequalities. The extremals for these inequalities are Blaschke factors. Thus one might explore Theorem 1.8.5 in several variables as a device that could give a soft construction of functions that will play the role of Blaschke factors in the function theory of several complex variables.

Recently, X. Huang has pushed the boundary rigidity ideas explored here even further. We give a quick indication of some of his ideas. Further details appear in [**HUA1**]-[**HUA3**].

THEOREM 1.8.7. *Let $\Omega \subset\subset \mathbb{C}^n$ be a smoothly bounded domain that satisfies Bell's Condition R ([**BEL1**]). Let $p \in \partial\Omega$ and $\sigma \in Aut\,(\Omega)$. If*

$$\sigma(z) = z + o(|z - p|)$$

then $\sigma \equiv id$.

This is a version of Cartan's uniqueness theorem at the boundary (see [**KRA1**], Ch. 11). It generalizes a result of [**KRA6**]. We shall discuss the proof of a version of this theorem in Section 1.10.

Now, in the vein of Theorem 1.8.6, Huang has proved the following (see [**HUA2**]).

THEOREM 1.8.8. *Let $\Omega \subset\subset \mathbb{C}^n$ be smoothly bounded, convex, and of finite type. Let $p \in \partial\Omega$. Then there exists an $m > 0$, depending on the local geometry of $\partial\Omega$ near p, such that, if f is holomorphic and*

$$f(z) = z + o(|z - p|^m) \qquad \text{as } z \to p,$$

then $f(z) \equiv z$.

The proof of this result is striking in that the choice of m depends on the exponent in the construction of a bounded plurisubharmonic exhaustion function. That is, the Diederich/Fornæss construction ([**DIF2**]) has the form $u = h \cdot (-\rho)^\tau$, where ρ is a defining function for the domain. Huang's exponent m depends on the Diederich-Fornæss exponent τ. The idea of Huang's proof is to first show that f must equal the identity on some holomorphic curve (in fact a complex geodesic of the Kobayashi metric). Then one studies the eigenvalues of Jac f restricted to this curve—much in the spirit of the Carathéodory-Cartan-Kaup-Wu theorem—and applies the Hopf lemma. Details are omitted.

However we note in passing that all of Huang's techniques depend on knowing some boundary regularity of complex geodesics for the Kobayashi metric. This issue came up implicitly in our study of Fatou theorems (in finding an invariant manner in which to define approach regions) and elsewhere in this book.

1.9. The Geometric Theory of Boundary Rigidity of Holomorphic Mappings. We begin this section with a return to the original question (from Section 1.8) about order of contact of boundary images of self-maps of domains. Assuming that our intuition in Section 1.8 was correct, and given that the analytic version of our question has the same answer in both one and several complex variables, then the issue is to use geometric analysis to separate the one variable situation from the several variable situation. We need to make our heuristic notion of order of contact, in the geometric sense, more precise. We begin with a definition:

DEFINITION 1.9.1. Let $\Omega \subset\subset \mathbb{C}^n$ be a smoothly bounded strongly pseudoconvex domain and fix $P \in \Omega$. Let ρ be a defining function for Ω near the boundary point P. Let Φ be a holomorphic mapping of Ω into Ω which continues C^N to a boundary neighborhood of P. We say that $\Phi(\partial\Omega)$ *has geometric order of contact N with $\partial\Omega$ at P* if $\Phi(P) = P$ and there is a positive C^k function h near P with the following properties: We normalize the coordinate system at P so that $z = 0$, the z_1 direction is the complex normal direction, and $z' = (z_2, \ldots, z_n)$ is the complex tangential direction. Then we require that

$$\rho \circ \Phi(z) = h \cdot \rho(z) + W(z),$$

where

$$|W(z)| \leq C \cdot \left(|z'|^4 + |z_1|^2\right)^{N/4}. \tag{1}$$

Any expression W that satisfies (1) will be called a term of *weight* N and will sometimes be denoted by W_N.

Now we have

PROPOSITION 1.9.2. *Let $\Omega \subseteq \mathbb{C}^n, n > 1$, be a strongly pseudoconvex domain and $P = 0$ a point of $\partial\Omega$. Assume that $\partial\Omega$ is C^6 near P. Let $\Phi : \Omega \to \Omega$ be a holomorphic mapping that continues C^6 to a boundary neighborhood P. Assume that $\Phi(\partial\Omega)$ has geometric order of contact 6 with $\partial\Omega$ at P and (as a normalization) that $\nabla\Phi(0) = id$. Then*

$$\Phi(z) = z + \mathcal{O}|z - P|^4.$$

In particular, it follows from Theorem 1.8.6 that Φ must be the identity mapping.

Remark: Notice that we do not assert the proposition for $n = 1$. In fact the Riemann Mapping Theorem shows that, for Ω the unit disc in $\mathbb{C}$, any order of contact may be achieved by non-trivial maps.

In fact the hypothesis of geometric order of contact 6 in this proposition is not the sharp one. Careful examination of the calculations shows that the final result will have a non-isotropic hypothesis. But by giving up some precision at this stage we are able to exploit the formalism of Chern and Moser [**CHM**] to give an argument that has some geometric structure. ■

It may be worth recording here a method for creating (close to) sharp examples in two dimensions in higher that suggest what the sharp form of the geometric contact theorem ought to be. For convenience we restrict attention to two dimensions and consider the unbounded realization $\mathcal{U} = \{(z, w) : \operatorname{Im} w > |z|^2\}$. Let ϕ be a holomorphic mapping of the upper half plane $\mathcal{H}$ (in one complex variable) to itself that takes 0 to 0 and takes the boundary into the boundary near 0 (i.e. is proper near 0). Assume that ϕ is one-to-one so that ϕ' never vanishes. Then we may consider the principal branch of $[\phi']^{1/2}$ and we may define

$$\Phi(z, w) = \left([\phi'(w)]^{1/2} z, \phi(w)\right).$$

Recall that the Schwarz lemma for the upper half plane says that

$$\frac{|\phi'(w)|^2}{|\operatorname{Im} \phi(w)|^2} \leq \frac{1}{|\operatorname{Im} w|^2}.$$

As a result, one may check that Φ maps $\mathcal{U}$ to itself.

We take the function ϕ to have the form

$$\phi(w) = w + \lambda w^3 + \cdots$$

near 0. It is easy to check that λ must be real in order for $\mathcal{H}$ to be preserved by ϕ. The fact that ϕ preserves $\partial\mathcal{H}$ near 0 implies that in fact

$$\operatorname{Im} \phi(w) = h(w, \bar{w}) \cdot \operatorname{Im} w$$

near 0, with h positive. In fact one may calculate by hand (writing $w = u + iv$) that

$$h(w, \bar{w}) = 1 + \lambda(3u^2 - v^2) + \cdots .$$

In view of this information about ϕ, we may write out Φ as

$$\Phi(z, w) = \left(z + \frac{3}{2}\lambda w^2 z + W_6, w + \lambda w^3 + W_8 \right).$$

The pullback of the defining function $\rho(z, w) = \operatorname{Im} w - |z|^2$ under Φ will thus have the form

$$\begin{aligned} \Phi^*(\rho) &= \operatorname{Im}(w + \lambda w^3) - |z|^2 - 3\lambda|z|^2 \operatorname{Re} w^2 + W_7 \\ &= h(w, \bar{w})\big(\operatorname{Im} w - |z|^2\big) + \big[\lambda(3u^2 - v^2) - 3\lambda \operatorname{Re} w^2\big]|z|^2 + W_7 \\ &= h(w, \bar{w})\big(\operatorname{Im} w - |z|^2\big) + 2\lambda v^2 |z|^2 + W_7. \end{aligned}$$

Notice that the obstruction to higher order of contact appears explicitly: the terms $2\lambda v^2|z|^2$ has weight 6 and serves this purpose. Notice that no such term can occur in the one complex variable situation. Further notice that the presence of the factor h gives us the flexibility to understand what is going on both in one and several variables. This completes the discussion of our example.

Proof of the Proposition: The proof of this implication is a rather difficult calculation. Therefore it seems appropriate to sketch the proof in the case when Ω is the ball in $\mathbb{C}^2$ and the point P is $\mathbb{1} = (1, 0)$. Since a strongly pseudoconvex point is biholomorphically the ball to fourth order (see [**FEF**]), the proof for a strongly pseudoconvex domain is formally the same. We work with the unbounded realization which we write as $\mathcal{U} = \mathcal{U}_{n+1} = \{(z, w) \in \mathbb{C}^2 : \operatorname{Im} w > |z|^2\}$. Of course $\rho(z, w) = \operatorname{Im} w - |z|^2$ is a defining function for $\mathcal{U}$. Our hypothesis is that

$$\rho \circ \Phi = h \cdot \rho + W_N, \tag{2}$$

where

$$|W_N| = |W_N(z, w)| \le C \cdot (|z|^4 + |w|^2)^{N/4}.$$

In what follows, the notation W_j will denote *any* expression satisfying

$$|W_j| \le C \cdot (|z|^4 + |w|^2)^{j/4}.$$

Without loss of generality we may make the following normalizations:

$$\Phi(z, w) = \big(z + A(z, w), w + B(z, w)\big),$$

where

$$\begin{aligned} A(z, w) &= \sum_{\nu=2}^{N} a_\nu(z, w) + W_{N+1} \\ B(z, w) &= \sum_{\nu=2}^{N} b_\nu(z, w) + W_{N+1}. \end{aligned}$$

Here a_ν, b_ν are homogeneous functions of degree ν in the dilation structure given by $\delta_\rho(z,w) = (\rho^2 z, \rho w), \rho \in \mathbb{R}^+$.

Normalizing h, we may write

$$h = 1 + \sum_{\nu=1}^{N} h_\nu + W_{N+1}.$$

Let a_{ν_0} be the first non-vanishing δ_ρ-homogeneous monomial in A. Then it is easy to see from (2) that b_{ν_0+1} is the first non-vanishing monomial in B. Thus

$$\begin{aligned}\Phi(z,w) &= \big(z + a_{\nu_0} + (\text{higher order terms}), \\ &\qquad w + b_{\nu_0+1} + (\text{higher order terms})\big)\end{aligned} \tag{3}$$

with $\nu_0 \geq 2$. Thus we see that the homogeneous of degree ν_0+1 part of equation (2) is

$$h_{\nu_0-1} \cdot (\operatorname{Im} w - |z|^2) = 2\operatorname{Re}\langle a_{\nu_0}, z\rangle + \operatorname{Im} b_{\nu_0+1}. \tag*{$(2)_{\nu_0+1}$}$$

[Here we assume, as we may, that $\nu_0 < N$.]

Now the operation

$$(a_{\nu_0}, b_{\nu_0+1}) \mapsto -2\operatorname{Re}\langle a_{\nu_0}, z\rangle + \operatorname{Im} b_{\nu_0+1},$$

evaluated at $\operatorname{Im} w = |z|^2$, is the operator "$L$" on formal power series that was defined in formula (2.6) of [**CHM**]. Equation $(2)_{\nu_0+1}$ says precisely that $L(a_{\nu_0}, b_{\nu_0+1}) = 0$. We shall consider $\nu_0 = 2, 3$ directly; the cases $\nu_0 = 4, 5$, which we shall also need, will then follow from Lemma 2.1 of [**CHM**].

The case $\nu_0 = 2$: In this case we have

$$\begin{aligned} h_1 &= \langle \alpha, z\rangle + \langle \bar\alpha, \bar z\rangle \quad , && \text{some } \alpha \in \mathbb{C}^n \\ a_2 &= Q(z) + \beta w \quad , && Q \text{ quadratic in } z,\ \beta \in \mathbb{C}^n \\ b_3 &= C(z) + w\langle z, \gamma\rangle \quad , && C \text{ cubic}, \gamma \in \mathbb{C}^n. \end{aligned}$$

Then equation $(2)_3$ becomes

$$\begin{aligned} &\left(\frac{\langle \alpha, z\rangle w}{2i} - \frac{\langle \bar\alpha, \bar z\rangle \bar w}{2i}\right) - \left(\langle \alpha, z\rangle |z|^2 + \langle \bar\alpha, \bar z\rangle |z|^2\right) + \left(\frac{\langle \bar\alpha, \bar z\rangle w}{2i} - \frac{\langle \alpha, z\rangle \bar w}{2i}\right) \\ &\quad = -\left(\langle Q(z), z\rangle + \overline{\langle Q(z), z\rangle}\right) - \left(w\langle \beta, z\rangle + \overline{w\langle \beta, z\rangle}\right) \\ &\qquad + \left(\frac{C(z)}{2i} - \frac{\overline{C(z)}}{2i}\right) + \left(\frac{w\langle z, \gamma\rangle}{2i} - \frac{\overline{w\langle z, \gamma\rangle}}{2i}\right). \end{aligned}$$

Comparing like monomials (labeled with indices (p, q)), we find that

$$\begin{array}{llcll} (\mathbf{3}, \mathbf{0}) & 0 &=& \frac{C(z)}{2i}, & \text{or } \ C(z) = 0\,; \\ (\mathbf{2}, \mathbf{0}) & \frac{w\langle z, \alpha\rangle}{2i} &=& \frac{w\langle z, \gamma\rangle}{2i}, & \text{or } \ \alpha = \gamma\,; \\ (\mathbf{2}, \mathbf{1}) & -\langle z, \alpha\rangle\langle z, z\rangle &=& -\langle Q(z), z\rangle, & \text{or } \ Q(z) = \langle z, \alpha\rangle z\,; \\ (\mathbf{1}, \mathbf{1}) & \frac{\langle \alpha, z\rangle w}{2i} - \frac{\langle z, \alpha\rangle \bar w}{2i} &=& -w\langle \beta, z\rangle - \bar w\overline{\langle \beta, z\rangle}\,. & \end{array}$$

The last line implies that $\alpha = -2i\beta$, some $\beta \in \mathbb{C}^n$.

Before proceeding, we recall that the automorphisms of $\mathcal{U}_{n+1}$ with $0 \in \partial\mathcal{U}_{n+1}$ a fixed point and normalized as in (3) have the form

$$(z,w) \mapsto \left(\frac{z + w\beta}{1 - 2i\langle z,\beta\rangle + cw}, \frac{w}{1 - 2i\langle z,\beta\rangle + cw}\right),$$

where $\beta \in \mathbb{C}^n$ is arbitrary and $\operatorname{Im} c = -|\beta|^2$. The calculations performed above tell us that, after composing with one of these automorphisms of $\mathcal{U}_{n+1}$, we may assume that (3) holds with $\nu_0 \geq 3$.

The case $\nu_0 = 3$: Now we have

$$\begin{aligned} h_2 &= q(z) + \alpha w + \bar{\alpha}\bar{w} + \bar{q}(z) \quad , && q \text{ quadratic}, \alpha \in \mathbb{C}^n; \\ a_3 &= C(z) + wL(z) \quad , && C \text{ cubic}, L \text{ linear}; \\ b_4 &= Q(z) + wS(z) + \gamma w^2 \quad , && Q \text{ quartic}, S \text{ quadratic}, j \in \mathbb{C}. \end{aligned}$$

Thus $(2)_4$ becomes

$$\begin{aligned} &\operatorname{Im}(qw + \alpha w^2) - 2\operatorname{Re} q|z|^2 - \operatorname{Im}(q\bar{w}) - 2\operatorname{Re}(\alpha w|z|^2) - |w|^2 \operatorname{Im}\alpha \\ &= \operatorname{Im}(Q + WS + \gamma w^2) - 2\operatorname{Re}\left[\langle C(z), z\rangle + w\langle L(z), z\rangle\right]. \end{aligned}$$

Comparing like terms now yields

$$\begin{array}{llcl} (\mathbf{4},\mathbf{0}) & Q & \equiv & 0; \\ (\mathbf{3},\mathbf{0}) & S & \equiv & q; \\ (\mathbf{2},\mathbf{0}) & \alpha & = & \gamma; \\ (\mathbf{3},\mathbf{1}) & C(z) & = & q(z)\cdot z; \\ (\mathbf{1},\mathbf{1}) & \operatorname{Im}\alpha & = & 0; \\ (\mathbf{2},\mathbf{1}) & -\alpha w\langle z,z\rangle - \frac{q\bar{w}}{2i} & = & -w\langle L(z), z\rangle \\ & & & \text{hence } q \equiv 0 \text{ and } L(z) = \alpha z. \end{array}$$

Note that we have now that $q = S = 0$ and $C = z \cdot q = 0$. The upshot of our calculation is that the mapping Φ agrees in form (up to order ν_0) with an automorphism of $\mathcal{U}$ that fixes 0. Composing Φ with the inverse of said automorphism (as we are obviously free to do), we have that Φ satisfies (3) with $\nu_0 = 4$.

The case $\nu_0 = 4, 5$: Applying Lemma 2.1 of [**CHM**], p. 233], we find that

$$(a_4, b_5) = (0,0) \qquad \text{and} \qquad (a_5, b_6) = 0.$$

Thus we have proved that geometric order of contact 6 implies that

$$\Phi(z,w) = (z,w) + (\text{terms of homogeneity at least six}).$$

This is the desired result. ∎

Again, details of these results in the strongly pseudoconvex case appear in [**BUK**]. Only a few remarks are made in [**BUK**] about the weakly pseudoconvex case. Meanwhile, X. Huang has initiated an intensive study of these boundary rigidity phenomena on some weakly pseudoconvex domains. In that setting, the

Lempert theory of extremal discs for the Kobayashi metric is unavailable. Therefore he has had to develop new techniques. We presented some of his results in Section 1.8.

1.10. Automorphism Groups of Domains. The results of the previous two sections paint a rather broad spectrum of boundary rigidity results; their genesis, however, springs from consideration of the rather more classical setting of automorphism groups of domains.

DEFINITION 1.10.1. Let $\Omega \subseteq \mathbb{C}^n$ be a domain. Let $Q \in \partial\Omega$. We say that Q is an *orbit accumulation point* if there is a point $P \in \Omega$ and automorphisms $\phi_j \subseteq \text{Aut}\,(\Omega)$ such that $\phi_j(P) \to Q$.

We note explicitly that the topology on the automorphism group is that of uniform convergence on compact sets (that is, normal convergence). This meshes naturally with the theory of normal families.

The automorphism group (group of biholomorphic self-maps, equipped with the compact-open topology) of the disc is generated by the *rotations*

$$\zeta \mapsto e^{i\alpha}\zeta$$

and the *Möbius transformations*

$$\phi_a : \zeta \mapsto \frac{\zeta - a}{1 - \bar{a}\zeta}$$

with $|a| < 1$. Observe that Aut (D) is non-compact because the mappings

$$\zeta \mapsto \frac{\zeta - (1 - 1/j)}{1 - (1 - 1/j)\zeta}$$

form a sequence of automorphisms that has no subsequence *converging to an element of* Aut (D).

By contrast, the annulus $\mathcal{A} \equiv \{\zeta : 1/2 < |\zeta| < 2\}$ does have compact automorphism group. The group is generated by the rotations and by the inversion $\zeta \mapsto 1/\zeta$. Thus, topologically, Aut $(\mathcal{A})$ is two copies of the circle and is compact. In particular, since the automorphism group of a domain is a biholomorphic invariant, we see (by an extremely artificial method) that the disc and the annulus are inequivalent.

Which bounded, planar domains in $\mathbb{C}$ have non-compact automorphism group? It is known that if the connectivity is finite, and at least two, then the automorphism group must in fact be a finite group. This can be seen using the uniformization theorem, or by mapping the region to a slit domain. If the connectivity is infinite, then things become more complicated:

EXAMPLE 1.10.2. Let $D \subseteq \mathbb{C}$ be the unit disc. Let $\phi(\zeta) = (\zeta - 1/2)/(1 - \zeta/2)$. Set $S = \bar{D}(0, 1/4)$. Finally, define

$$\Omega = D \setminus \bigcup_{j=-\infty}^{\infty} \phi^j(S).$$

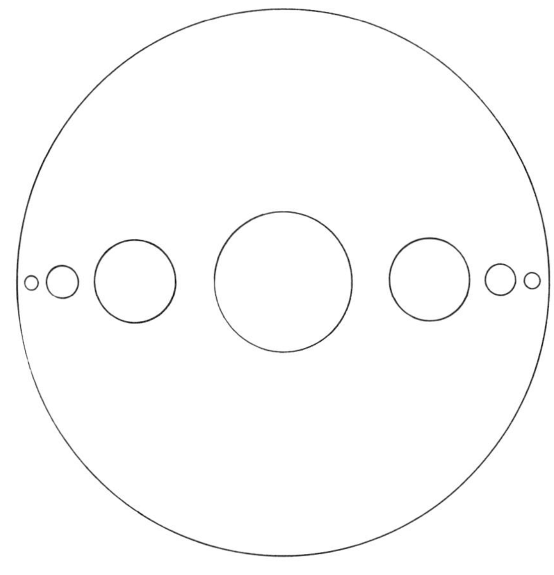

FIGURE 1.22

Refer to Figure 1.22.

Then the automorphism group of Ω contains $\{\phi^j\}$, which is non-compact. Notice that, by hyperbolic geometry, if the removed discs were configured differently then the automorphism group could change drastically. In particular, it could become finite or trivial.

In this example, the automorphism group is a one-parameter group, but elaboration of this construction can produce more complicated automorphism groups. I do not know which non-compact groups can arise as the automorphism groups of planar domains (with no further restrictions imposed on the domain). It is also unknown which finite groups can arise as automorphism groups of planar domains. The famous upper bound, due to Hurwitz, of $84(g-1)$ on the order of the automorphism group of a compact Riemann surface in terms of the genus is essentially sharp. Using the notion of the double of a plane domain, one can get some bounds on the size of the automorphism group of a bounded, planar domain. But sharp results of this nature are unknown.

It was Andreotti who first noticed that the Chern number could be used to obtain some bounds on the order of the automorphism group in higher dimensions. Some interesting partial results also appear in [**HUC**] and [**XIG**]. The results in the latter paper, while not completely general, are fairly sharp. There is also the following remarkable theorem of Bedford/Dadok [**BEDD**] and Saerens/Zame [**SAZ1**], which says that all groups arise:

THEOREM 1.10.3. *Let G be any compact Lie group. Then there is a strongly pseudoconvex domain Ω in some $\mathbb{C}^n$ such that $Aut(\Omega) = G$.*

In fact the domain Ω can be taken to have real analytic boundary (as follows

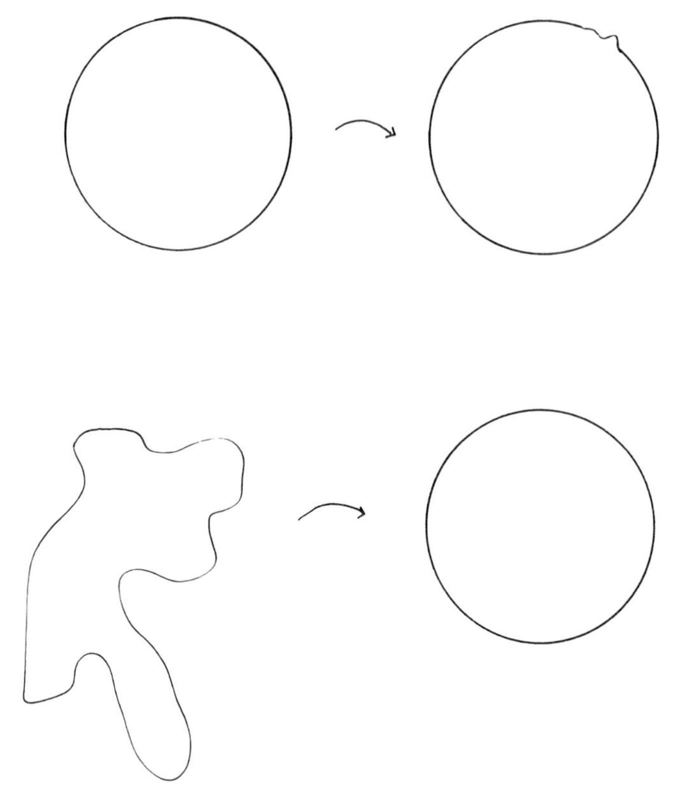

FIGURE 1.23

from work of Greene and Krantz [**GK5**]), and to be arbitrarily close to the ball. The dimension n depends on the complexity of the group G. In fact, in the Bedford/Dadok proof, one constructs the domain Ω in the representation space for G. The strategy is to begin with, say, the ball B, and to add bumps and ridges that shave away unwanted parts of $\text{Aut}\,(B)$ and leave only the symmetries in G. The difficulty is, for example, in distinguishing groups with the same orbits—such as $O(n)$ and $SO(n)$. It is at this juncture that the theory of irreducible unitary representations plays a useful role.

The Saerens/Zame proof is a transversality argument: one constructs a large moduli space and takes a slice.

We next state a theorem from [**GK5**] and connect it with the last result by way of a conjecture. The theorem is motivated by considering classical Euclidean geometry of the plane. Consider the symmetries provided by rigid motions. Observe (Figure 1.23) that if a planar figure has a great deal of symmetry then an arbitrarily small perturbation of that figure can destroy that symmetry. However if a figure lacks symmetry then it requires a perturbation of at least a certain size to create symmetry. Here is a precise formulation of this notion in the several complex variables setting:

THEOREM 1.10.4 (GREENE/KRANTZ). *Let Ω_0 be a strongly pseudoconvex domain. Let $G = Aut(\Omega_0)$. If Ω is a sufficiently small, smooth perturbation of Ω_0 then $Aut(\Omega)$ is a subgroup of G in a natural way. In fact there is a diffeomorphism (generally not a biholomorphism) $\Psi : \Omega \to \Omega_0$ such that the mapping*

$$Aut(\Omega) \ni \phi \mapsto \Psi \circ \phi \circ \Psi^{-1} \in G$$

is a univalent homomorphism of groups.

Note that we could not hope that Ψ is a biholomorphic mapping: first, because then the automorphism groups would be isomorphic but, second, because two different strongly pseudoconvex domains are generically biholomorphically inequivalent. The theorem is known (by different techniques—see [**GK4**]) to hold in the C^2 topology, but without the conclusion about the diffeomorphism Φ. It is known to fail in the $C^{1-\epsilon}$ topology.

Now we conjecture the following refinement of the theorem of Bedford/Dadok and of Saerens/Zame:

> **CONJECTURE:** Let $\Omega_0 \subseteq \mathbb{C}^n$ be a smoothly bounded, strongly pseudoconvex domain and G its automorphism group. Let H be a subgroup of G. Then there is a strongly pseudoconvex domain Ω, arbitrarily close to Ω_0 in a smooth topology, such that $\text{Aut}(\Omega) = H$.

Next, for motivation of our discussions of automorphism groups, we consider the following one variable result. It follows naturally from the uniformization theorem, but we shall attack its proof more directly.

PROPOSITION 1.10.5 ([**KRA10**]). *Let $\Omega \subset\subset \mathbb{C}$ be a domain with smooth boundary (requiring that the boundary be locally the graph of a Lipschitz function will suffice). Assume that $Aut(\Omega)$ has non-compact automorphism group. Then Ω is (conformally equivalent to) the disc.*

SKETCH OF PROOF: Using a normal families argument and the open mapping principle, one sees that non-compactness of $\text{Aut}(\Omega)$ implies that there are a point $P \in \Omega$, a point $Q \in \partial\Omega$ and elements $\psi_j \in \text{Aut}(\Omega)$ such that $\psi_j(P) \to Q$. [We shall treat this point in detail below.]

Now let f be a function, continuous on $\bar{\Omega}$ and holomorphic on Ω, such that $f(Q) = 1$ and $|f(\zeta)| < 1$ for all $\zeta \in \bar{\Omega} \setminus \{Q\}$. Such an f is easily constructed by conformally mapping Ω to the disc (we use here the smoothness of the boundary in a mild way) so that Q goes to $1 \in \partial D$ and postcomposing that conformal mapping with the function $\mu(\zeta) = (\zeta + 1)/2$.

The family $\{f \circ \psi_j\}$ is normal. If $f \circ \psi_{j_k} \to g$ normally, then it must be that both $|g| \leq 1$ and $g(P) = 1$. These assertions are consistent only if $g \equiv 1$. But then $f \circ \psi_{j_k} \to 1$ normally, and this means that $\{\psi_{j_k}\}$ converges to the constant function Q uniformly on compact subsets of Ω. [We remark here that a small extra argument will show that the full sequence $\{\psi_j\}$ converges to the constant function Q uniformly on compacta.]

Now we could argue as follows: if $\gamma : [0,1] \to \Omega$ is a loop in Ω, $\gamma(0) = \gamma(1)$, then its image G is a compact set in Ω. Let U be a neighborhood of Q such that $U \cap \Omega$ is topologically trivial (*this* is where we really use the boundary smoothness). By the preceding paragraph, there is a k so large that $\phi_{j_k}(G) \subseteq U \cap \Omega$. It follows that $\phi_{j_k}(G)$ may be deformed to a point in $U \cap \Omega$. But then G may be deformed to a point in Ω. Thus γ is homotopically trivial. We conclude that Ω is simply connected and the Riemann mapping theorem finishes the proof.

There is some interest in avoiding the use of the Riemann mapping theorem, since such a powerful tool is unavailable in the function theory of several complex variables. Thus we briefly describe another method for treating the last step. Write $U \cap \Omega$ as the increasing union of sets $V_1 \subset\subset V_2 \subset\subset \cdots$ such that $\cup_m V_m = U \cap \Omega$ and so that each V_m is topologically trivial. Then for each j a number $\ell = \ell(m)$ may be chosen such that

- $\ell(1) < \ell(2) < \cdots$;
- $\left(\psi_{j_{k_{\ell(m)}}}\right)^{-1}(V_m) \subset\subset \left(\psi j_{k_{\ell(m+1)}}\right)^{-1}(V_{m+1})$, each ℓ;
- $\bigcup_m \left(\psi_{j_{k_{\ell(m)}}}\right)^{-1}(V_m) = \Omega$.

It follows that Ω is biholomorphic to the disc, because one may pass to a normal subsequence of the (suitably normalized) conformal mappings $\left(\psi_{j_{k_{\ell(m)}}}\right)^{-1} : V_m \to D$ to exhibit this biholomorphism explicitly.

This discussion completes the proof of the proposition. ■

It was Bun Wong who first proved a generalization of the last proposition to several complex variables. Namely, the only strongly pseudoconvex domain with non-compact automorphism group is the ball. A self-contained proof of this assertion is discussed later in this section. Another, more geometric way to look at this theorem is as follows (this approach may be found in the work of Klembeck [**KLE**]). By non-compactness, there are automorphisms ψ_j of Ω and a point $Q \in \partial\Omega$ such that if $P \in \Omega$ then $\psi_j(P) \to Q$ (we shall say more about this important consideration below). Consider the holomorphic sectional curvature of the Bergman metric at P. By invariance, this tensor is just the same as the curvature tensor for the metric at $\psi_j(P) \equiv P_j$. But $P_j \to Q \in \partial\Omega$. One may use the Fefferman asymptotic expansion for the Bergman kernel near Q to get asymptotic expressions for the Bergman metric, and hence for its curvature, near Q. The upshot is that the curvature tensor is arbitrarily close to the constant curvature tensor of the ball, provided only that j is large enough. But this means that the curvature tensor at P is arbitrarily close to that of the ball. Thus the curvature tensor at P must *equal* the constant curvature tensor of the ball. But P was arbitrary. Thus the Bergman curvature on all of Ω is constant. But a theorem of Lu Qi-Keng [**LUQ**] then implies that Ω is biholomorphic to the ball. That is what we wished to prove.

In fact one need not assume that the whole domain is strongly pseudoconvex in order to make the Bun Wong theorem valid. Only points near the boundary orbit accumulation point need be such. This crucial observation is due to Rosay, and we shall say more about it in what follows.

The paper [**GK7**] offers yet another proof of the Bun Wong-Rosay theorem using an invariant of Bergman geometry that involves two fewer derivatives than Bergman metric curvature. An invariant with fewer derivatives has the attractive feature that it is more susceptible of generalization to a broader class of domains.

We have spent some time in this book touting the problem of classifying those smoothly bounded domains with non-compact automorphism group. For completeness and perspective, we should note explicitly that the classical domains of Cartan (see [**HEL**] and [**GK1**]) have non-compact, indeed *transitive* automorphism groups. However, with the exception of the ball, *these domains do not have smooth boundaries.* The classification of *all* domains with non-compact automorphism group, including both non-smooth domains and unbounded domains, appears to be quite difficult. It would be of some interest in this context to consider domains having self-similar or *fractal* boundaries. Such domains would perforce have boundaries that are not smooth. Thus far, no work has been done on this topic. On the other hand, considerable progress has been made in classifying certain homogeneous complex manifolds. See [**HUL**] for some definitive results.

The following elegant result may be found in [**NAR**] or [**GK1**]. It is a nice application of the open mapping principle. We note for the record that the result is susceptible of a number of generalizations—indeed the automorphisms may be replace by "approximate automorphisms" (see [**GK6**]). The result that we now state is fundamental to the way that we think about domains with non-compact automorphism group; it has been used twice already in our discussion of the Bun Wong-Rosay theorem.

PROPOSITION 1.10.6. *Let $\phi_j \in Aut(\Omega)$. Assume that $\{\phi_j\}$ converges normally (uniformly on compact sets) to a limit mapping ψ. Then either $\psi \in Aut(\Omega)$ or the image of ψ lies in $\partial\Omega$.*

In fact something slightly stronger is true; for the sake of the proof that we shall sketch, it is useful to have the more general statement.

THEOREM 1.10.7 (CARTAN). *Let $\phi_j \in Aut(\Omega)$. Assume that $\{\phi_j\}$ converges normally (uniformly on compact sets) to a limit mapping $\psi : \Omega \to \mathbb{C}^n$. Then the following three properties are equivalent:*

1: $\psi \in Aut(\Omega)$;

2: $\psi(\Omega) \not\subset \partial\Omega$;

3: *There exists a point $P \in \Omega$ such that the Jacobian matrix $Jac_{\mathbb{C}}\psi(P)$ has non-zero determinant.*

SKETCH OF PROOF: That (1) $\Rightarrow$ (2) is obvious.

For (1) $\Rightarrow$ (3), notice that the fact that ψ is invertible certainly implies that the Jacobian at each point is invertible. So the Jacobian determinant at each point will be non-zero.

If we assume (3), then the holomorphic inverse function theorem (see [**KRA1**]) implies that P has a neighborhood on which ψ acts diffeomorphically. Thus certainly $\psi(\Omega) \not\subset \partial\Omega$, which is assertion (2).

Now assume (2) and we seek to prove (3). We choose a relatively compact open set $U \subseteq \Omega$ that gets mapped to a relatively compact open $V \subseteq \Omega$ by ψ. Shrinking these sets slightly, we may assume that ϕ_j maps U into V for j large. But then one can see that that a subsequence $\phi_{j_k}^{-1}$ converges on a relatively compact open subset of V to a limit mapping μ. One then argues, by taking limits, that $\psi \circ \mu$ is the identity on V. In particular, the Jacobian matrix of the limit mapping ψ will be non-degenerate at points of U. That establishes (3).

For (3) $\Rightarrow$ (1), notice that $m_j(z) \equiv \det \mathrm{Jac}_{\mathbb{C}}\phi_j(z)$ is non-vanishing on all of Ω for each j. By a suitable version of Hurwitz's theorem (see [**KRA1**]), $\det \mathrm{Jac}_{\mathbb{C}}\psi$ never vanishes. Thus ψ is an open mapping. In particular, the image of ψ lies entirely in Ω. Using arguments that we have seen in (2) $\Rightarrow$ (3), we may see that a subsequence of $\{\phi_j^{-1}\}$ converges to a mapping τ that has non-vanishing Jacobian determinant on $\psi(\Omega)$. But then τ is open as well. Again, as in (2) $\Rightarrow$ (3), one sees that $\psi \circ \tau = \mathrm{id}$ on an open set and likewise for $\tau \circ \psi$. It follows that ψ is an automorphism. ∎

EXAMPLE 1.10.8. Let $\Omega = D$, the unit disc in $\mathbb{C}$. Define

$$\phi_j(\zeta) = \frac{\zeta + (1 - 1/j)}{1 + (1 - 1/j)\zeta}.$$

These are automorphisms of Ω, and they converge normally to the constant mapping $\psi(\zeta) \equiv 1$.

On the other hand, the mappings

$$\mu_j(\zeta) = \frac{\zeta + 1/j}{1 + (1/j)\zeta}$$

converge to the identity mapping.

Thus, in this simple context, we have exhibited the dialectic of the theorem.

Notice also that the automorphism group of an annulus in the plane, consisting as it does of rotations and inversion, is perforce compact. And plainly no point has an orbit that escapes to the boundary. ∎

Some interesting refinements of Cartan's theorem appear in [**BEL2**].

The following elegant result, which we have already used, is a corollary of Cartan's theorem:

PROPOSITION 1.10.9. *Let $\Omega \subset\subset \mathbb{C}^n$. Then $\mathrm{Aut}(\Omega)$ is non-compact if and only if there are a point $P \in \Omega$, a point $Q \in \partial\Omega$, and automorphisms $\phi_j \in \mathrm{Aut}(\Omega)$ such that $\phi_j(P) \to Q$.*

SKETCH OF PROOF: Assume that there are points P and Q as in the proposition. Then any limit mapping of the ϕ_j will map Ω (at least in part) into the boundary. Thus the limit mapping will not be an automorphism and Aut (Ω) is not compact.

If instead there are no such P and Q then fix some $P \in \Omega$. Then $\{\phi(P) : \phi \in \Omega\}$ lies in a relatively compact subset of Ω. Thus we are in case (2) of Cartan's theorem and any limit function must be an automorphism. It results that Aut (Ω) is compact. ■

From our point of view, the result that opened up the modern work in the area of automorphism groups is that of Bun Wong [**WONG**]. We first need to enunciate Bun Wong's lemma. We make the following definition:

DEFINITION 1.10.10. Let $\Omega \subseteq \mathbb{C}^n$ be open. Define the *Carathéodory volume form* on Ω by

$$M_\Omega^C(z) = \sup\{|\det f'(z)| : f \in B(\Omega), f(z) = 0\}.$$

[Here $B(\Omega)$ denotes the set of all holomorphic mappings of Ω to B.]

Define the *Kobayashi-Eisenman volume form* on Ω by

$$M_\Omega^K(z) = \inf\{1/|\det f'(0)| : f \in \Omega(B), f(0) = z\}.$$

[Here $\Omega(B)$ denotes the set of all holomorphic mappings of B to Ω.]

LEMMA 1.10.11 (BUN WONG). *Let $\Omega \subset\subset \mathbb{C}^n$ be a domain. Suppose that there is a point $P \in \Omega$ such that $M_\Omega^C(P) = M_\Omega^K(P)$. Then Ω is biholomorphic to the ball.*

SKETCH OF PROOF: In case Ω is a "nice" domain (that is, if Ω is taut—in particular, if Ω is a smoothly bounded domain of holomorphy) then there is a limit mapping $\phi : B \to \Omega$ that realizes $M_\Omega^K(P)$. There is always a limit mapping $\psi : \Omega \to B$ that realizes $M_\Omega^C(P)$ because B is "nice". But then our hypothesis implies that $F = \psi \circ \phi : B \to B$ has the property that $F(0) = 0$ and det $\mathrm{Jac}_{\mathbb{C}} F(0)$ has modulus 1. By a version of the Schwarz lemma, F must be a rotation (see also Cartan's theorem in Section 1.7). It follows that ϕ, ψ are biholomorphisms, and hence that Ω is biholomorphic to B.

The case in which Ω is not assumed to be nice is more technical. Details may be found in [**KRA1**]. ■

Now we have Bun Wong's original theorem:

THEOREM 1.10.12 (BUN WONG). *Let $\Omega \subseteq \mathbb{C}^n$ be strongly pseudoconvex with C^2 boundary. If Ω has non-compact automorphism group then Ω is biholomorphic to the unit ball.*

SKETCH OF PROOF: By 1.10.9, there are $P \in \Omega$ and $\phi_j \in \mathrm{Aut}\,(\Omega)$ be such that $\phi_j(P) \to Q \in \partial\Omega$. It is well known (see [**KRA1**]) that there is a peaking function f on Ω for the point Q. By considering, $f \circ \phi_j$, and applying Hurwitz's theorem, one can see that $\phi_j \to Q$ *uniformly on compact subsets of* Ω. But then, by a normal families argument, one sees that if U is a neighborhood of P in $\mathbb{C}^n$ then $M_C^\Omega(P) = M_C^\Omega(\phi_j(P)) \cdot |\det \phi_j'|$ is arbitrarily close to $M_C^{\Omega\cap U}(\phi_j(P))$ for j large enough. A somewhat different normal families argument shows that $M_K^{\Omega}(P)$ is arbitrarily close to $M_K^{\Omega\cap U}(\phi_j(P)) \cdot |\det \phi_j'|$ when j is large enough.

But, on a small enough neighborhood U, $\Omega \cap U$ can be approximated closely by $B \cap U$, where B is the unit ball. Here we use a careful analysis of the positivity condition that defines strong pseudoconvexity. In particular,

positive definiteness $\sim$ diagonalizability $\sim$ all eigenvalues equal $\sim$ ball.

Thus

$$\frac{M_K^{\Omega}(P)}{M_C^{\Omega}(P)} \sim \frac{M_K^{\Omega\cap U}(\phi_j(P)}{M_C^{\Omega\cap U}(\phi_j(P)} \sim \frac{M_K^{B\cap U}(\phi_j(P))}{M_C^{B\cap U}(\phi_j(P))} \sim 1.$$

Passing to the limit, we find that $\frac{M_K^{\Omega}(P)}{M_C^{\Omega}(P)} = 1$. Using the remarks preceding the proof, we find that Ω is biholomorphic to the ball. ■

Using additional normal families arguments, Rosay [**ROS**] was able to prove the following essentially deeper theorem:

THEOREM 1.10.13 (ROSAY). *Let $\Omega \subseteq \mathbb{C}^n$ be an arbitrary bounded domain. Assume that $Q \in \partial\Omega$ is strongly pseudoconvex, in particular that the boundary near Q is C^2 and strongly pseudoconvex. If there exists a point $P \in \Omega$ and automorphisms ϕ_j such that $\phi_j(P) \to Q$ then Ω is biholomorphic to the ball.*

Rosay's proof is similar to the first one sketched above, but he had to work harder for lack of global knowledge about Ω.

The line of reasoning raised by these two theorems is this: if a boundary point Q is an orbit accumulation point of the action of the automorphism group on the domain Ω, then to what extent is the global geometry of Ω determined by the local (Levi) geometry of Q? For example, in the Rosay theorem, the point Q is an orbit accumulation point and the Levi form is positive definite at Q. Using elementary algebra—that a positive definite Hermitian matrix can be diagonalized, etc.—one may see that Q is very nearly like the ball. This is true in the sense that there are local holomorphic coordinates near Q such that a local defining function may be written as

$$\rho(z) = -1 + |z|^2 + \mathcal{O}|z - \mathbb{1}|^4.$$

The conclusion of Rosay's theorem is that in fact Ω is *globally* biholomorphic to the ball. What other domains, besides the ball, may serve as models? What other geometric properties can the automorphism group action detect? Here are some sample theorems:

PROPOSITION 1.10.14. *Let $\Omega \subseteq \mathbb{C}^n$ be a domain. Let Q be a C^2 boundary point which is also an orbit accumulation point. Then Q is is a point of pseudoconvexity.*

PROOF. Suppose not. Then the Levi form at Q has a negative eigenvalue. That means that any holomorphic function on Ω (or even on a neighborhood U

of Q in Ω) continues analytically to a full neighborhood V of Q. In particular, *all automorphisms of* Ω *continue to a fixed full neighborhood* V *of* Q.

Let $P \in \Omega, \phi_j \in \text{Aut}(\Omega)$, and $\phi_j(P) \to Q \in \partial\Omega$. Let $P_j = \phi_j(P)$. By the Cauchy estimates, we see from the last paragraph that $|d\phi_j^{-1}(P_j)|$ are bounded, uniformly in j. Thus $|d\phi_j(P)|$ are bounded from zero. Likewise, the second derivatives of the ϕ_j are uniformly bounded above on a neighborhood of P. If B is a small open Euclidean ball centered at P then it follows that the balls $\phi_j(B)$ each contain a Euclidean ball of uniform size. But this contradicts the hypothesis that $\phi_j(P) \to Q \in \partial\Omega$. Thus Q must be a point of pseudoconvexity. That completes the proof. ∎

Proofs that are similar to, but more technical in their execution than, the one just given suffice to prove the following two variants. In these statements, $\hat{K}$ denotes the hull of K with respect to the family of holomorphic functions on Ω. In 1.10.15 and 1.10.16 no boundary smoothness of Ω is assumed.

PROPOSITION 1.10.15. *Let* Ω *be a bounded domain (not necessarily pseudoconvex) in* $\mathbb{C}^n$. *Let* $p \in \Omega$ *and* $K \subset\subset \Omega$. *If a sequence* $\phi_j \in \text{Aut}(\Omega)$ *has the properties that* $\phi_j(p) \in \hat{K}$ *for each* j *and that* $\lim_{j\to\infty} \phi_j(p)$ *exists, then* $\lim_{j\to\infty} \phi_j(p)$ *belongs to* Ω *and not to* $\partial\Omega$.

PROPOSITION 1.10.16. *Let* $\Omega \subseteq \mathbb{C}^n$ *be a bounded domain (not necessarily pseudoconvex) and* $K \subset\subset \Omega$. *Fix* $p_0 \in \partial\Omega$. *Then there is no sequence* $\{\phi_j(p)\}$, $p \in \Omega$, $\phi_j \in \text{Aut}(\Omega)$, *with the property that: for each open polydisc* D *centered at* p_0, *some component of* $D \cap \Omega$ *contains a point of* $\hat{K}$ *and a point* $\phi_j(p)$.

To put these more technical results in perspective, the reader should bear this in mind: the property of *not* being a domain of holomorphy means that some $K \subset\subset \Omega$ has the property that $\hat{K}$ is *not* relatively compact in Ω. Thus $\hat{K}$ must escape to the boundary somewhere. These propositions assert that the point of escape cannot be an orbit accumulation point.

DEFINITION 1.10.17. Let Q be a boundary point of a domain $\Omega \subseteq \mathbb{C}^n$. We say that Q is *variety free* if there is no germ V of a complex analytic variety such that $P \in V \subseteq \partial\Omega$.

Of course a finite type point in any dimension is variety free. It is a theorem of Bedford/Fornæss [**BEF**] that a finite type point in the boundary of a domain in dimension 2 is in fact a peak point for the algebra $A(\Omega) = C(\bar{\Omega}) \cap$ holomorphic functions. [Other very interesting proofs of this result appear in [**FOS**], [**FOM**].] Possession of a peak point is a strictly stronger condition than being variety free.

PROPOSITION 1.10.18. *Let* $\Omega \subset\subset \mathbb{C}^n$, *assume that* Ω *has non-compact automorphism group, and further suppose that* $Q \in \partial\Omega$ *is an orbit accumulation point and that* Q *is variety free. If the boundary of* Ω *is* C^1 *near* Q *then* Ω *is topologically trivial.*

PROOF. By hypothesis there is a point $P \in \Omega$, a boundary point Q, and automorphisms ϕ_j such that $\phi_j(P) \to Q$. By a minor variant of the peaking function argument that we have used several times, one sees that ϕ_j converges, uniformly on compact sets, to the constant mapping Q. Let U be a neighborhood of Q such that $U \cap \Omega$ is topologically trivial (this is the only point at which we use the hypothesis that $\partial\Omega$ is C^1). Let $K_1 \subset\subset K_2 \subset\subset \cdots \subset\subset U \cap \Omega$ be topologically trivial compact sets which exhaust $U \cap \Omega$ monotonically. It is an elementary exercise in topology to see that, for $k_1 << k_2 << \cdots$ large enough, the sets $(\phi_{k_j})^{-1}(K_j)$ exhaust Ω. Each of these is topologically trivial, and it follows that so is Ω. ■

PROPOSITION 1.10.19. *Let $\Omega \subset\subset \mathbb{C}^n$, assume that Ω has non-compact automorphism group, and further suppose that $Q \in \partial\Omega$ is an orbit accumulation point, that Q has a boundary neighborhood that is pseudoconvex, and that Q is variety free. Then Ω is pseudoconvex.*

PROOF. Similar to the last proposition. ■

Note that it is necessary in the last proposition that we assume that Ω be pseudoconvex in an entire boundary neighborhood of Q. Examples of Bedford/Pinchuk below (see [**BEP3**]) show that the result is false if we only suppose that Q alone is pseudoconvex.

It is natural at this point to look at some examples to help to understand the limitations of the last three results. Example 1.10.2 applies to show us that, in the absence of boundary smoothness, conclusions like that of the Bun Wong-Rosay theorem do not apply. This example was given in dimension one. We also have:

EXAMPLE 1.10.20. Let $\Omega = B(0,1) \subseteq \mathbb{C}^2$ and $S = \bar{B}(0,1/4)$. Define the mapping

$$\Phi(z_1, z_2) = \left(\frac{z_1 - 1/2}{1 - (1/2)z_1}, \frac{(\sqrt{3}/2)z_2}{1 - (1/2)z_1} \right).$$

Then we may produce the region

$$\Omega = B(0,1) \setminus \bigcup_{j=-\infty}^{\infty} \Phi^j(S).$$

The automorphism group of Ω contains $\{\phi^j\}$, hence is *not compact.* But Ω is *not* topologically trivial, and is *not* a domain of holomorphy (hence is *not* pseudoconvex).

Of course the domain Ω is also not smoothly bounded. ■

EXAMPLE 1.10.21 (BEDFORD/PINCHUK). We work in $\mathbb{C}^n$. Let $m_j > 0$ be an integer for each $j = 1, \ldots, n$. Set $\delta_j = 1/(2m_j)$. If $J = (j_1, \ldots, j_n)$ and $K = (k_1, \ldots, k_n)$ are multi-indices, then we set

$$\begin{aligned} \mathrm{wt}(z^J) &= j_1\delta_1 + \cdots + j_n\delta_n \\ \mathrm{wt}(z^J \bar{z}^K) &= \mathrm{wt}(z^J) + \mathrm{wt}(\bar{z}^K). \end{aligned}$$

Consider a real-valued polynomial $p(z_1, \ldots, z_n, \bar{z}_1, \ldots, \bar{z}_n)$ of the form

$$p(z, \bar{z}) = \sum_{\mathrm{wt}(z^J)=\mathrm{st}(\bar{z}^K)=1/2} a_{JK} z^J \bar{z}^K .$$

Of course the hypothesis of being real valued implies that $a_{JK} = \bar{a}_{KJ}$.

Define the domain

$$\Omega = \Omega_p = \{(w, z_1, \ldots, z_n) \in \mathbb{C} \times \mathbb{C}^n : |w| + p(z, \bar{z}) < 1\}.$$

Define the mapping $(w, z) \mapsto (w^*, z^*)$ by

$$w = \left(1 - \frac{iw^*}{4}\right)\left(1 + \frac{iw^*}{4}\right)^{-1}, \quad z_j = z_j^* \left(1 + \frac{iw^*}{4}\right)^{-2\delta_j}, j = 1, \ldots, n.$$

This transformation maps Ω onto an unbounded domain

$$\Omega^* = \{(w^*, z_1^*, \ldots, z_n^*) \in \mathbb{C} \times \mathbb{C}^n : \mathrm{Im}\, w^* + p(z^*, \bar{z}^*) < 0\}.$$

Notice that this unbounded realization Ω^* has defining function that is independent of $\mathrm{Re}\, w$. As a result, the automorphism group contains all translations in the $\mathrm{Re}\, w$ coordinate: in particular, the automorphism group is non-compact. On the other hand, if p is not plurisubharmonic then the domain Ω will not be pseudoconvex.

Even when Ω is not pseudoconvex, the orbit accumulation point is pseudoconvex. But this point does not have a boundary neighborhood of pseudoconvex points. ■

This interesting example demonstrates the limitations, and the sharpness, of our results to the effect that orbit accumulation points must be pseudoconvex and that (bounded) domains with non-compact automorphism group and boundary orbit accumulation point having a *neighborhood* of pseudoconvex points must be globally pseudoconvex.

Which smoothly bounded pseudoconvex domains have non-compact automorphism group? This is a natural question, in view of what has gone before. In two complex dimensions, the examples that spring to mind are the so-called "egg domains:"

$$E_m = \{(z_1, z_2) : |z_1|^2 + |z_2|^{2m} < 1\}.$$

The domain E_m has the automorphisms

$$\Phi_j(z_1, z_2) = \left(\frac{z_1 - a}{1 - (a)z_1}, \frac{(1 - |a|^2)^{1/2m}}{(1 - (a)z_1)^{1/m}}\right),$$

for any complex number a such that $|a| < 1$. These plainly form a non-compact set in the $\mathrm{Aut}\,(E_m)$ topology (just let $|a| \to 1$). That these are the "only" domains in $\mathbb{C}^2$ with non-compact automorphism group is a beautiful theorem of Bedford/Pinchuk [**BEP1**]:

THEOREM 1.10.22 (BEDFORD/PINCHUK). *Let $\Omega \subset\subset \mathbb{C}^2$ have real analytic boundary. Assume that Ω has non-compact automorphism group. Then Ω is biholomorphic to one of the domains E_m.*

The Bedford/Pinchuk theorem has been generalized ([**CAT5**]) to domains of finite type. Thus we have a complete classification of finite type domains in dimension two with non-compact automorphism group.

In higher dimensions, there are domains other than eggs that have non-compact automorphism group. For example, the domain

$$\Omega = \{(z_1, z_2, z_3) \in \mathbb{C}^3 : |z_1|^2 + (|z_2|^2 + |z_3|^2)^{2m} < 1\} \tag{1}$$

has non-compact automorphism group but (as can be seen by an inspection of weakly pseudoconvex boundary points) is not biholomorphic to an egg. A plausible conjecture is that a domain with non-compact automorphism group should (properly, holomorphically) cover the ball (as do the egg domains). However again the example (1) shows that this is not the case.

A more likely conjecture, due to Catlin and Pinchuk, is based on an analysis of the following type of domain: Let $\mathbf{m} = (m_1, \ldots, m_n)$ be an n-tuple of positive integers. Assume that $m_1 = 1$. Say that a multi-index $(\alpha_1, \ldots, \alpha_n)$ is of **m**-*type* if

$$\sum_j \frac{\alpha_j}{m_j} = 1.$$

Define

$$E_{\mathbf{m}} = \{z \in \mathbb{C}^n : \sum_{\alpha \text{ of } \mathbf{m}-type} |z^\alpha|^2 < 1\}.$$

Then we have:

> **CONJECTURE (Catlin/Pinchuk):** Let $\Omega \subset\subset \mathbb{C}^n$ be a pseudoconvex domain with smooth boundary. If Ω has non-compact automorphism group then Ω is biholomorphic to a domain $E_{\mathbf{m}}$ as above.

This conjecture has been confirmed by Bedford/Pinchuk [**BEP3**] when Ω satisfies the additional hypothesis that it is of finite type in $\mathbb{C}^2$. However in the more recent paper [**BEP3**], Bedford and Pinchuk have observed that it is too restrictive to suppose that a domain with non-compact automorphism group is Reinhardt. In fact Example 1.10.21 suggests that the correct conjecture may be that the defining function, in suitable coordinates, is "balanced" in a sense given by non-isotropic weights. We leave it for the reader to infer the correct conjecture from that example or to refer to [**BEP3**].

The preceding discussion, especially the positive results in the papers of Bedford and Pinchuk, have led to the following companion conjecture:

> **CONJECTURE (Greene/Krantz):** If Ω is smoothly bounded and $Q \in \partial\Omega$ is an orbit accumulation point then Ω is a point of finite type. In particular, the point Q is variety free.

The point of the second conjecture is that one need not *assume* in advance that the domain (or at least the orbit accumulation point) is pseudoconvex.

In the case that Ω is convex then K. T. Kim has (in effect) verified the second conjecture in [**KIM5**]. As a part of this work, he has shown that a domain with a piecewise Levi flat boundary and noncompact automorphism group is in fact a product domain. Related results also appear in [**GK7**]. Results that support the conjecture, in the case of domains with circular symmetry, can be found in [**SUN**].

Kim has proved a number of theorems, using the scaling technique that amplify the theme that the geometry of a boundary orbit accumulation point determines the geometry of the entire domain. One limitation of this interesting work is that the orbit accumulation point must be "convexifiable"—this is necessary for the scaling argument to converge. See [**KIM1**]-[**KIM5**]. Frankel ([**FRA1**], [**FRA2**]) also has some interesting work in this connection.

Akio Kodama and his collaborators ([**KOD1**] - [**KOD4**] and [**KKM**] for instance) have developed interesting techniques for characterizing domains with orbit accumulation points that are locally biholomorphic to ellipsoids and generalized ellipsoids.

Given the evidence that we have at this writing, it appears that new ideas will be required to treat non-convexifiable domains with non-compact automorphism groups.

1.11. Uniform Estimates for Automorphisms. We would like to use this section to explicate a powerful analytic tool that was first discovered and exploited in [**GK5**]. As befits the theme of this book, it was discovered in the pursuit of geometric goals. But the tool is one of hard analysis. Additional references are [**GK5**] and [**BED2**].

THEOREM 1.11.1. *Let $\Omega \subset\subset \mathbb{C}^n$ be a smoothly bounded domain that satisfies condition R. Let $G \subseteq Aut(\Omega)$ be a compact subset. Let $0 < M \in \mathbb{Z}$. Then there is a constant $C_M > 0$ such that*

$$\sup_{\phi \in G} \sup_{z \in \Omega} \max_{|\alpha| \leq M} \left| \left(\frac{\partial}{\partial z} \right)^{\alpha} \phi(z) \right| \leq C_M .$$

Greene and Krantz first proved this result for strongly pseudoconvex domains by an indirect argument using Bergman representative coordinates (discussed below). Barrett [**BAR**], Bell [**BEL2**], and Bedford [**BED2**] have constructed arguments that are perhaps more natural, and which apply in the greater generality of Condition R domains (which includes all finite type domains). Bedford [**BED2**] has proved even more: that the action $\bar{\Omega} \times G \to \bar{\Omega}$ is C^∞.

Two trivial examples illustrate what is being said here. The disc $D \subseteq \mathbb{C}$ has non-compact automorphism group, as we have noted several times before. And the derivatives (say the first derivatives, for specificity) of automorphisms are *not* uniformly bounded: the specific automorphisms

$$\phi_j(\zeta) = \frac{\zeta - (1 - 1/j)}{1 - (1 - 1/j)\zeta}$$

explicitly illustrate this point. By contrast, an annulus *does* have compact automorphism group generated by the rotations and reflection. It is plain that the derivatives of these automorphisms *do* enjoy a uniform bound.

In short, this theorem goes beyond the classical results of Fefferman [**FEF**], Bell [**BEL2**], and others in that it makes statement about *uniform smoothness*, up to the boundary, of an entire family of biholomorphic mappings.

The result is best seen by contradiction. If it fails, then there are automorphisms ϕ_j for which the left side blows up. By compactness of G, one obtains a subsequence that converges uniformly on compact subsets to an automorphism Φ_0. By Bell [**BEL2**], using Condition R together with the usual machinery for smoothness to the boundary for biholomorphic mappings, the automorphisms must in fact converge in the $C^M(\bar{\Omega})$ topology. This leads to a contradiction.

An alternative proof—in effect the one used in [**GK5**]—is instructive to consider. It also allows us to showcase the concept of *Bergman representative coordinates.* We begin with a brief discussion of this idea.

One of Bergman's dreams was to find a "canonical imbedding into space" of any given domain in $\mathbb{C}^n$. In one complex variable, the disc is the canonical imbedding of any simply connected domain (not the plane). For a multiply connected domain in the plane, the canonical realization could be a slit domain (there are other possibilities as well). In an effort to find these canonical imbeddings in several variables, Bergman invented the representative coordinate concept.

Let $\Omega \subseteq \mathbb{C}^n$ be a bounded domain and $K(z, \zeta) = K_\Omega(z, \zeta)$ its Bergman kernel. Fix a point $w \in \Omega$. Then the representative coordinates centered at w are given by the function

$$z \longmapsto \frac{\partial}{\partial \bar{\zeta}_j} \log \frac{K(z,\zeta)}{K(\zeta,\zeta)}\bigg|_{\zeta = w} \quad , \quad j = 1, \ldots, n.$$

The following two important properties are elementary to check (or see Section 2.4 for details):

1: For z in a neighborhood of w the representative coordinate mapping is univalent. Hence it really does provide a local coordinate system near w.

2: Any biholomorphism of Ω becomes linear when it is written in representative coordinates.

Details of these two assertions are provided in Section 2.4 below.

The rub is that the neighborhood on which Bergman's mapping gives a valid coordinate system is in general indeterminate: it can be rather small (compared to Ω), indeed it is often relatively compact in Ω. It is known (see [**GK2**]) that for domains sufficiently near the ball the representative coordinates provide a *global coordinate system* on the closure of the domain. This provides a new way to think about Fefferman's theorem that biholomorphic mappings extend smoothly to the closure. It also provides a method for proving a striking equivariant imbedding theorem for domains. This latter result is very much in the spirit of Bergman's "canonical imbedding" program. [For the sake of clarity here we omit some technical details from the statement; refer to [**GK2**], [**GK3**] for the precise result.]

THEOREM 1.11.2 (GREENE/KRANTZ). *Let $\Omega \subseteq \mathbb{C}^n$ be a smoothly bounded domain which is sufficiently near the ball (in a smooth topology on domains—see Section 1.6 for a discussion of this concept). If Ω is not biholomorphic to the ball, then there is an imbedding $\Phi : \Omega \to \mathbb{C}^n$ such that the automorphism group of $\Phi(\Omega)$ acts on $\Phi(\Omega)$ by restriction of a compact group of unitary matrices (acting on $\mathbb{C}^n$ in the usual way) to Ω.*

The theorem provides a way to circularize Ω, much in the spirit of the Riemann mapping theorem in one complex variable. In fact the imbedding Φ is obtained as follows. One first proves, using an old idea of Hadamard, that the automorphism group of the domain has a fixed point P (see Section 1.7). Then the (globally defined) representative coordinate system associated to the point P realizes each biholomorphism as a complex linear map that fixes P (which is the origin in the new coordinates). That observation, together with the functoriality of the construction, guarantees that each automorphism is now unitary.

Return for the moment to Theorem 1.11.1. To obtain a slightly different proof, suppose that the theorem is not true. Then, after passing to suitable subsequences, we find a sequence $\{\phi_j\}$ of automorphisms and points P_j at which some derivative D of order not exceeding M satisfies $|D\phi_j(P_j)|$ blows up. It follows from Cauchy estimates that the P_j cannot accumulate at an interior point of Ω. Let Q be a boundary accumulation point of the $\{P_j\}$. A detailed analysis of Fefferman's asymptotic expansion for the Bergman kernel, together with difficult calculations, shows that the Bergman representative coordinates at an *interior point sufficiently near* P are in fact valid *up to* P. But then one can look at the ϕ_j in these special coordinates and see that compactness has been violated.

We now present some results that illustrate the power of the uniform estimates on compact sets of automorphisms (we mention in passing that similar techniques may be used to obtain uniform estimates on derivatives of compact sets of mappings between *different* domains).

We will use an analytic tool that should be of considerable independent interest. We begin with some background. Part of the interest of the Poincaré metric on the disc is that it turns the disc into a complete Kähler manifold. At the same time, when one wants to explore the interaction of boundary analysis on the circle and interior analysis on the disc, then it is not useful to work with the Poincaré metric because the boundary is infinitely far away. A similar situation holds with any of the standard invariant metrics—Bergman, Carathéodory, Kobayashi, or Sibony—on a strongly pseudoconvex domain.

If one want to study the interplay between boundary analysis and interior analysis, then it is best to have an *invariant* metric that is *smooth on the closure of* Ω. If the group $\mathrm{Aut}\,(\Omega)$ is non-compact, then it is impossible to have such a metric. For, by Proposition 1.10.9, there would have to be an interior point P, a boundary point Q, and automorphisms ϕ_j such that $\phi_j(P) \to Q$. But then, using the mean value theorem, one can see that the metric could not be smooth at the boundary point Q.

On the other hand, for a compact automorphism group there is hope (think of the automorphism group of the annulus as an example). In fact we have the following:

THEOREM 1.11.3. *Let $\Omega \subseteq \mathbb{C}^n$ be a smoothly bounded domain that satisfies Bell's Condition R. Let G be a compact group of automorphisms of Ω. Then there is a Hermitian metric h, smooth on $\bar{\Omega}$, that is invariant under the action of G.*

PROOF. Let $g = \{g_{ij}\}$ be any smooth metric on $\bar{\Omega}$—say the Euclidean metric. If $\xi \in \mathbb{C}^n$ then define

$$\|\xi\|_h \equiv \int_{\tau \in G} \|\tau_* \xi\| \, d\tau. \qquad (1)$$

Because of the uniform estimates on derivatives of elements of G (see Theorem 1.11.1 above), we may differentiate under the integral sign in (1) and see that the entries h_{ij} of h are smooth on $\bar{\Omega}$.

Now it follows from the construction of h that any automorphism of Ω is an isometry of the metric h. ■

The theorem bears some discussion. Recall (Section 1.10) that a strongly pseudoconvex domain has non-compact automorphism group if and only if it is the ball. Thus the theorem applies to the full automorphism group of any strongly pseudoconvex domain save one. It is an open problem (see Section 1.10) to determine which weakly pseudoconvex domains have non-compact automorphism groups. But, as far as we can tell, the case of compact automorphism group is generic, *even if one restricts attention to topologically trivially domains.* Thus the theorem has wide applicability.

As a first simple application of the new invariant metric (which, by the way, is not unique since it depends on the initial choice of g) we have

THEOREM 1.11.4. *Let Φ be an automorphism of a strongly pseudoconvex domain Ω. Assume that Ω is not biholomorphic to the ball. If $P \in \partial\Omega$, $\Phi(P) = P$, and $\Phi'(P) = id$, then $\Phi(z) \equiv z$.*

Before sketching the proof, we note that Fefferman's theorem [**FEF**] implies that Φ will continue smoothly to the boundary, hence it makes sense to consider the expressions $\Phi(P)$ and $\Phi'(P)$. The theorem is false if $\mathrm{Aut}\,(\Omega)$ is non-compact (in particular if Ω is the ball). For instance, let $\Omega = \{z \in \mathbb{C} : \mathrm{Im}\, z > 0\}$ (which is biholomorphic to the unit disc). Set $\Phi(z) = z/(z+1)$. Then $\Phi(0) = 0, \Phi'(0) = 1$, yet clearly $\Phi \neq \mathrm{id}$. X. Huang [**HUA1**] has explored generalizations of the theorem, and has considered more thoroughly how the failure of the theorem is linked to non-compactness of the automorphism group.

PROOF OF THEOREM 1.11.3: We use the smooth-across-the-boundary metric of Theorem 1.11.4. It follows from the construction of h that any automorphism of Ω is an isometry of the metric h. Recall that an isometry is completely determined by its first order behavior at a point: the only isometry that fixes P and has derivative id at P is the identity. This can be seen by noticing that a geodesic emanating from P is uniquely determined by its tangent vector at P. The hypothesis about first order behavior guarantees that these geodesics are preserved by the isometry, at least locally. Thus the isometry is locally the identity. By unique continuation, it is the identity globally.

The hypothesis of the theorem asserts that the automorphism Φ has the same first order behavior at P as the identity mapping. Both maps are isometries of the metric h. Thus they must coincide. That concludes the proof. ■

Next we turn to a deep result that is closely related to the boundary geometry of Chern and Moser. A (special case of) a very striking result of Kruzhilin and Vitushkin [**VIT2**] is as follows:

THEOREM 1.11.5 ([**VIT2**]). *Let $\Omega \subset\subset \mathbb{C}^n$ be a strongly pseudoconvex domain with real analytic boundary. Assume that Ω is not biholomorphic to the ball. Then there is a neighborhood U of $\bar{\Omega}$ such that every element $\Phi \in \mathrm{Aut}\,(\Omega)$ analytically continues to U.*

It is instructive to first consider this result in one complex dimension. Let us assume for simplicity that $\Omega \subseteq \mathbb{C}$ is smoothly bounded. The hypothesis that $\mathrm{Aut}\,(\Omega)$ is compact means that Ω is not simply connected. Take, for example, Ω to be an annulus with center the origin and radii 2, 4. Then it easy to see that $\mathrm{Aut}\,(\Omega)$ is generated by rotations and inversion. In particular, the common domain of analytic continuation for automorphisms may be taken to be $U = \{z : 1 < |z| < \infty\}$. Generically, the automorphism group of a domain with k holes will (canonically) be a subgroup of the semi-direct product of k groups that are isomorphic to the automorphism group of an annulus (see [**HEI1**]). Thus it is not difficult to see that U exists. An alternative approach would be to use the uniformization theorem.

In several variables, one has to work harder. Vitushkin et al [**VIT1**], [**VIT2**] proved a local result about germs of mappings of strongly pseudoconvex hypersurfaces with a fixed point. The size of U in their proof depended on measuring the "distance of the hypersurface" from being spherical, in the sense of CR geometry. They used a variant of the Chern-Moser invariants to perform their measurements. (Note that this work is deep. Recently V. Ezhov [**EZH**] has constructed an example in $\mathbb{C}^{13}$ of a strongly pseudoconvex CR hypersurface containing a point P such that the isotropy group of P cannot be linearized.)

An alternate proof may be obtained by exploiting Bergman representative coordinates (see [**FUKS**]], [**KRA6**], [**GK2**], [**GK3**], [**GK5**]). If $\partial\Omega$ is real analytic and strongly pseudoconvex then it is known (see [**BEL3**]) that, for $z \in \Omega$ fixed, $K(z, \cdot)$ is real analytic up to the boundary. Thus, with some calculations using the uniform estimates 1.11.1 above, one may arrange for the Bergman representative coordinates at P sufficiently near the boundary to be valid, as real analytic coordinate charts, across the boundary. In particular, these charts, which linearize any given automorphism, give an explicit device for arranging the desired analytic extension to a pre-determined patch. Details appear in [**KRA6**].

At first it seems that the Kruzhilin/Vitushkin result has the advantage of being local (as opposed to the global nature of the Greene/Krantz proof just outlined). But some localization techniques of [**SAZ2**] make it possible to localize the argument in [**KRA6**].

In fact there is a third, and more elementary, proof of the theorem that is due to Coupet [**COU**].

It is not entirely clear whether the analogue of this theorem holds for a compact group of automorphisms of a weakly pseudoconvex domain with real analytic boundary. While it is true (by results of [**BJT**], for example) that each biholomorphism will analytically continue to a neighborhood of the closure, the calculation of Bergman representative coordinates as in the proof in [**KRA6**] breaks down. Also, the analysis of chains as in [**VIT1**],[**VIT2**] fails. There is no evidence that the result itself should fail, but current techniques seem to be inadequate.

In the paper [**GK5**], the technique of uniformity of estimates on automorphism derivatives, and particularly of the smooth-across-the-boundary metric was considerably refined. In particular, by patching the metric h with the Bergman metric, by using an exhaustion argument, and by using the compactness of the automorphism group in a decisive way, it was possible to construct an invariant metric $\tilde{h}$ with the following properties: there is a relatively compact subdomain $U \subset\subset \Omega$ such that (i) $\bar{U}$ is invariant under $\mathrm{Aut}\,(\Omega)$; (ii) the only possible isometries of $\bar{U}$ are elements of $\mathrm{Aut}\,(\Omega)$ or their conjugates (this last property holds for the Bergman metric on a reasonable class of domains—see [**GK5**] and references therein). Thus it was possible to modify $\tilde{h}$ near $\partial\Omega$ so that the metric is a product metric near $\partial\Omega$.

The advantage of having a product structure near the boundary is that one may then consider the metric *double* $\hat{\Omega}$ of Ω. Every element of $\mathrm{Aut}\,(\Omega)$ extends in a natural way to $\hat{\Omega}$ and so does the metric. It can be argued that the only isometries of $\hat{\Omega}$ are the (extended) elements of Aut (Ω) and their conjugates.

Finally, Ebin [**EBI**] has a beautiful theorem about semi-continuity of isometry groups: Let M be a compact Riemannian manifold equipped with a metric σ. Let I_σ be the isometry group of this metric. Ebin's theorem is that if the metric is perturbed slightly in the smooth topology to a new metric σ' then the resulting isometry group $I_{\sigma'}$ is a subgroup of I_σ in a natural way: by conjugation by a diffeomorphism.

In [**GK5**] it was argued that Ebin's theorem could be applied to the action of $\text{Aut}\,(\Omega) \cup \overline{\text{Aut}\,(\Omega)}$ on $\hat{\Omega}$ (here the overline denotes *conjugation*). The resulting theorem is the semi-continuity of automorphisms group result that was discussed in Section 1.10.

While much of the machinery that goes into the proof of this theorem is valid on domains with Condition R and compact automorphism groups, there are certain crucial junctures where stability properties of the Bergman metric are needed. For a limited class of weakly pseudoconvex domains, some progress has been made on these stability properties in [**GEB**] and [**DIH**].

In [**GK10**], it was shown that the last theorem holds for perturbations in the C^2 topology. On the other hand, it was not possible to produce the conjugating map Φ that sends automorphisms of the perturbed domain univalently to automorphisms of the base domain (see the precise statement in Theorem 1.10.4). It was also shown that the results fail in the $C^{1-\epsilon}$ topology. These counterexamples were described in Section 1.10. It is not known what happens in topologies between C^1 and C^2.

Chapter 2. Function Spaces and Real Analysis Concepts

2.1. Bloch Spaces. There is an extensive literature of Bloch spaces in the function theory of one complex variable. Bloch spaces in several complex variables are less well developed. Our purpose in this section is to discuss some results in the several variable setting. Although generalizations to domains of finite type in $\mathbb{C}^2$ are now possible, we shall concentrate on the more familiar settings of the ball and strongly pseudoconvex domains.

Let $D \subseteq \mathbb{C}$ be the unit disc. We begin by recalling the classical theorem of Bloch (see [**HAY**], p. 170):

THEOREM 2.1.1 (A. BLOCH). *There is an absolute constant B such that the following is true: Let $f : D \to \mathbb{C}$ be a holomorphic function normalized so that $f(0) = 0$ and $f'(0) = 1$. Then the disc $D(0, B)$ is in the image of f, and univalently so.*

We call $D(0, B)$ a *Schlicht disc* in the image of f. There is great interest in determining the optimal value B of Bloch's constant given by the theorem. The classical estimates, primarily due to Ahlfors [**AHL2**] and refined by Heins [**HEI2**], Pommerenke [**POM2**], and Ahlfors/Grunsky [**AHG**], are these:

$$.433 < \frac{\sqrt{3}}{4} < B \leq \frac{1}{\sqrt{1+\sqrt{3}}} \frac{\Gamma(1/3)\Gamma(11/12)}{\Gamma(1/4)} < .4719.$$

Pommerenke [**POM2**] introduced the Bloch constant B_0 for the more restricted class of locally univalent holomorphic functions (no branch points). He proved the estimates

$$.5 < B_0 \leq \frac{\Gamma(1/3)\Gamma(5/6)}{\Gamma(1/6)} < .5433. \qquad (*)$$

The most interesting recent development in these matters is that Mario Bonk [**BONK**] has improved the lower bound for B by $+10^{-14}$.

In the 1980's it was discovered that Pommerenke's proof of the lower bound for B_0 was in error. On the other hand, in Spring 1992 H. Yanagihara [**YAN**] announced that he *could* prove the lower bound of .5; indeed, he improved it to $.5 + 10^{-47}$.

Fix $z_0 \in D$. Assume that $f : D \to \mathbb{C}$ satisfies $f(z_0) = 0$ and $f'(z_0) \neq 0$. If we apply the Bloch theorem to the function

$$g(z) = \frac{f((z + z_0)/(1 + \bar{z}_0 z)) - f(z_0)}{f'(z_0)(1 - |z_0|^2)}$$

(note that $g(0) = 0$ and $g'(0) = 1$), then we find that there is a Schlicht disc in the image of f, centered at $f(z_0)$, of radius at least $B|f'(z_0)|(1 - |z_0|^2)$. This expression leads to the classical definition of "Bloch function."

DEFINITION 2.1.2. A holomorphic function $f : D \to \mathbb{C}$ is said to be *Bloch*, or in the Bloch class, if

$$|f'(\zeta)| \leq C \cdot (1 - |\zeta|^2)^{-1}$$

for all $\zeta \in D$.

Notice that the definition puts an upper bound on the radii of Schlicht discs in the range of f. The condition in the definition arises naturally when one applies the Cauchy estimates to a bounded analytic function. However the Bloch class is larger than H^∞ : for example, the function $f(\zeta) = \log(1 - \zeta)$ is Bloch. Bloch functions are of interest in part because they share some of the important properties of H^∞ functions. In particular, a Bloch function is normal (see Section 2.2 and [**COL**]).

Curiously the Bloch space has no natural relationship with any of the standard classical function spaces (Bergman, H^p, etc.) except for H^∞. The boundary behavior of elements of the Bloch space is quite pathological. In fact it is possible for a Bloch function f to have almost everywhere radial limit function that is bounded but f is not in H^∞—indeed, f cannot be in any H^p class. It is also possible for a Bloch function to have radial limits almost nowhere: indeed, the normalization mapping from the disc to the $\mathbb{C} \setminus \{\mathbb{Z} + i\mathbb{Z}\}$ is such a mapping.

It is known that if a Bloch function f has radial boundary limits almost everywhere then the coefficients a_ν of its power series expansion must satisfy

$$\frac{1}{n^2} \sum_{\nu=1}^{n} |a_\nu|^2 \nu^2 \to 0$$

(see [**TIM1**], [**TIM2**]). This last condition is equivalent to

$$\iint\limits_{|z| \leq r} |f'(z)| \, dxdy = o\left(\frac{1}{1 - r}\right).$$

The papers [**TIM1**] and [**TIM2**] are good references for background and basic pathologies of the Bloch space.

The problem of boundary limits for Bloch functions is a delicate one, and is not understood at this time. On the other hand, since a Bloch function is normal (Section 2.2), it has many favorable properties regarding boundary behavior.

The Bloch space is a natural one to study: it is known, for instance (see [**RUT**]) that the Bloch space is the unique largest Möbius invariant space that possesses non-zero linear functionals. We mention in passing that there is also a unique minimal invariant space and a unique invariant Hilbert space (see [**PEL**] for more on these matters).

For our purposes, it is useful to give a more geometric definition of the Bloch space. This is consistent with other definitions that have been given in several complex variables, and makes the formulation of several of the properties of these spaces more elegant.

DEFINITION 2.1.3. Let $\Omega \subset\subset \mathbb{C}^n$ be a domain equipped with the Kobayashi metric F_K^Ω. Let $f : \Omega \to \mathbb{C}$ be a holomorphic function. We say that f is *Bloch*, or lies in the Bloch class, if

$$|f_*(P)\xi| \leq CF_K^\Omega(P, \xi)$$

for all $P \in \Omega$ and $\xi \in T_P(\Omega)$. There is no loss of generality to identity $T_P(\Omega)$ with a copy of $\mathbb{C}^n$ and thus to think of ξ as a vector in $\mathbb{C}^n$.

This definition is consistent, on reasonable domains such as strongly pseudoconvex and finite type in $\mathbb{C}^2$, with ostensibly different definitions given in other contexts (see [**HAH1**], [**TIM1**], [**TIM2**]). While the Bergman metric has been used in some of these other contexts, it turns out that the Kobayashi metric is most convenient for our purposes. On strongly pseudoconvex domains ([**GRA1**], [**FEF**]), finite type domains in $\mathbb{C}^2$ ([**CAT4**]), and convex domains (see [**MCN6**], [**YU2**], [**FRA2**]), all three of the standard canonical metrics are known to be comparable.

The first order of business is to see that our definition of Bloch function has certain standard backwards compatibility properties. We begin with a little notation.

If $P' \in \partial\Omega$ then we let $\nu_{P'}$ denote the unit outward normal at P'. There is a $\delta_0 > 0$ sufficiently small that if $P' \in \partial\Omega, 0 < \delta < \delta_0, \mathbf{b} = \delta\nu_{P'}$, and $a = P' - \delta\nu_P$, then

$$D(a, \mathbf{b}) \equiv a + D\mathbf{b} = \{a + \zeta\mathbf{b} : \zeta \in D\}$$

is called a *complex normal disc* for Ω.

Next, let $f : \Omega \to \mathbb{C}$ be a holomorphic function. A Euclidean disc

$$D(w, r) \equiv \{\zeta \in \mathbb{C} : |\zeta - w| < r\}$$

is called a *Schlicht disc* in the range of f if there exists a holomorphic $\phi : D \to \Omega$ (here D, unadorned, is the unit disc in the plane) such that $f \circ \phi$ on D is univalent onto $D(w, r)$. Notice that this definition, while consistent with that of Schlicht disc when the complex dimension n is one, is conceptually different when the dimension is at least two. For a holomorphic function from an open set of dimension two or greater into $\mathbb{C}$ is never one-to-one. Hence we have to use the parameterizing map ϕ in our definition of Schlicht disc.

Finally, we refer the reader to Section 2.2 for a discussion of the concept of normal function and family. In several complex variables the topic is more subtle than in one, for a sequence of mappings may degenerate without the mappings being compactly divergent (diverging to infinity). For the moment, we say that a family of holomorphic functions $\mathcal{F} = \{f_\alpha\}$ on Ω is *finitely normal* if every sequence in $\mathcal{F}$ has a subsequence which converges uniformly on compact subsets of Ω to a holomorphic function f.

THEOREM 2.1.4. *Let*

$$\Omega = \{z \in \mathbb{C}^n : \rho(z) < 0\}$$

be a strongly pseudoconvex domain. We assume that ρ is C^2 and that $\nabla\rho \neq 0$ on $\partial\Omega$. We let d_K be the integrated Kobayashi metric on Ω. The following are equivalent:

1: *f is a Bloch function.*

2: *The radii of Schlicht discs in the range of f are bounded above.*

3: *As a function from the metrized space (Ω, δ_K) to the metrized space $(\mathbb{C}^n$, Euclidean metric), we have that f is uniformly continuous.*

4: *We have*

$$\sup_{z\in\Omega} |\nabla_\nu f(z)||\rho(z)| < +\infty,$$

where ρ is any defining function for Ω and ∇_ν is the normal derivative.

5: *The family*

$$\left\{g(\zeta) \equiv f(\mathbf{a}+\zeta\mathbf{b}) - f(\mathbf{a}) : D \to \mathbb{C} \mid \mathbf{a} \in \mathbb{C}^n, \mathbf{b} \in \mathbb{C}^n, \mathbf{a} + D\mathbf{b} \in \Omega\right\}$$

is a family of Bloch functions on the disc with uniformly bounded Bloch norms (note that the discs being considered here are not necessarily normal).

6: *The family*

$$\left\{g(\zeta) \equiv f(\mathbf{a}+\zeta\mathbf{b}) - f(\mathbf{a}) : D \to \mathbb{C} \mid \mathbf{a} + D\mathbf{b} \text{ is a normal disc in } \Omega\right\}$$

is a family of Bloch functions on the disc with uniformly bounded Bloch norms.

7: *The family*

$$\mathcal{D} \equiv \left\{g(z) = f \circ \phi(\zeta) - f \circ \phi(0) \mid \phi : D \to \Omega \text{ holomorphic}\right\}$$

is a finitely normal family.

8: *The family in (7) is a family of Bloch functions with uniformly bounded Bloch norms.*

9: *There exists a number $\alpha > 0$ such that, for any disc*

$$D(\mathbf{a}, \mathbf{b}) \equiv \left\{\mathbf{a} + \zeta\mathbf{b} : \zeta \in D\right\} \subseteq \Omega$$

we have

$$\int_{t_0}^{t} \exp\left\{\frac{1}{\alpha} f(\mathbf{a}+\zeta\mathbf{b})\right\} d\zeta \neq 0$$

for all $t, t_0 \in D, t_0 \neq t$.

PROOF. We sketch the gist of the proof.

(1) $\Rightarrow$ **(3)** That f is Bloch means that ∇f is distance decreasing from the Kobayashi metric to the Euclidean metric. But then f must be distance decreasing in the integrated metrics.

(3) $\Rightarrow$ **(7)** By the distance decreasing property of the Kobayashi metric, we know that the family $\mathcal{D}$ is equicontinuous from the Poincaré (Kobayashi) metric to the Euclidean metric. Since the origin is pinned down—all functions in the family vanish at 0, we may conclude, by a variant of the Ascoli-Arzela theorem, that the family is finitely normal.

(7) $\Rightarrow$ **(2)** For all $\psi \in \mathcal{D}$ we know that $|\psi'(0)| \leq M$, some $M > 0$, else $\mathcal{D}$ would not be normal. Let $h : D \to \Omega$ be such that $f \circ h(\zeta) = w + r\zeta$ (this is just another way of saying that the disc $D(w, r)$ is a Schlicht disc in the image of f).

Set $g(\zeta) = f \circ h(\zeta) - f \circ h(0)$. Then $g \in \mathcal{D}$ hence $r = |g'(0)| \leq M$. But this just says that the radius of any Schlicht disc in the image of f is bounded above by M.

(2) $\Rightarrow$ **(8)** Obvious by contrapositives.

(8) $\Rightarrow$ **(5)** $\Rightarrow$ **(6)** Trivial by inspection.

(6) $\Rightarrow$ **(4)** Let $z \in \Omega$ be sufficiently near the boundary and let $\zeta \mapsto z + r\zeta\nu$ be a normal disc centered at z. Set $g(\zeta) = f(z + r\zeta\nu)$. If ρ is a defining function for Ω then $|\rho(z)| \leq Cr$. Also

$$|\rho(z)| \cdot |\nabla_\nu f(z)| \leq Cr \cdot |\nabla_\nu f(z)| = C|g'(0)|.$$

By (6), $\|g\|_{\mathcal{B}(D)} \leq M$, some $M > 0$. Since $|g'(0)| \leq \|g\|_{\mathcal{B}(D)}$, we conclude that

$$\rho(z)|\nabla_\nu f(z)| \leq C \cdot M.$$

That is what we wished to prove.

(4) $\Rightarrow$ **(1)** This is technically the most difficult step of the proof, but we shall only sketch the idea.

By standard arguments, it holds on any smoothly bounded domain that if

$$|\nabla_\nu f(z)| \leq C \cdot |\rho(z)|^{-1} \qquad (*)$$

then

$$|\nabla_\tau f(z)| \leq C \cdot |\rho(z)|^{-1/2} \qquad (**)$$

for any unit complex tangent vector τ. See [**TIM1**], [**KRA1**], and references therein. But now, if Ω is strongly pseudoconvex, standard estimates for the Kobayashi metric (see [**ALA1**], [**GRA1**]) imply that f is Bloch.

(5) $\Leftrightarrow$ **(9)** We shall omit this proof since it is not germane to the main ideas of this section. The argument involves a standard criterion for univalence related to the Schwarzian derivative (see [**POM1, POM2**]).

The proof of the theorem is complete. ■

We note in passing that the only part of the theorem that actually uses strong pseudoconvexity is (4) $\Rightarrow$ (1). For finite type domains in $\mathbb{C}^2$, estimates of Catlin [**CAT4**] provide the necessary tools to prove (4) $\Rightarrow$ (1). For more general domains the matter is open.

Our next focus of attention is the connection between the Bloch space and various BMO spaces. In order to facilitate the discussion, we introduce some additional notation.

Let us restrict attention to a strongly pseudoconvex domain Ω. Let dV be standard Euclidean/Lebesgue volume measure on Ω and define the volume element

$$d\lambda(z) = |\rho(z)|^{-n-1}dV(z).$$

Let d_K denote the integrated Kobayashi distance on Ω. We fix a number $R > 0$ and consider Kobayashi metric balls in Ω : if $P \in \Omega$ then

$$B_K(P, R) \equiv \{z \in \Omega : d_K(P, z) < R\}.$$

It is worth noting that, once the radius $R > 0$ is fixed then the Kobayashi metric ball $B_K(P, R)$ has Euclidean radius comparable to $1/\delta_\Omega(P)$ in normal directions and comparable to $1/\sqrt{\delta_\Omega(P)}$ in tangential directions. The constants of comparability of course depend on R.

We also mention that, were we to develop these ideas on a domain of finite type in $\mathbb{C}^2$ then the shape of Kobayashi metric balls near a point of type m would be $1/\delta$ in the normal directions and $1/\delta^{1/m}$ in tangential directions (this follows from the estimates in [**CAT4**], for instance). However there is a subtlety involved: whereas, by compactness, the points in a strongly pseudoconvex boundary can be taken to be "uniformly strongly pseudoconvex" (eigenvalues of the Levi form bounded from zero), such is not the case on a finite type boundary—in fact it is never the case. Even in the simple example

$$\Omega = \{(z_1, z_2) : |z_1|^2 + |z_2|^4 < 1\},$$

we can see that points $(z_1, z_2) \in \partial\Omega$ with $z_2 \neq 0$ are of type 2 (strongly pseudoconvex). Yet, as $z_2 \to 0$ the boundary point becomes type 4. This means that the coefficient of ∂r in the commutator $[L, \bar{L}]$ is becoming smaller and smaller as $z_2 \to 0$. The expression $\Lambda(r)$ (vis. τ) introduced in Sections 1.1, 1.3, 1.4 is used to control the relative sizes of the various commutators.

Return to the strongly pseudoconvex case. Using the metric balls β and the volume element dV we may define a solid BMO space by saying that $f \in BMO_{solid}(\Omega)$ if

$$\sup_{P\in\Omega} \inf_{c\in\mathbb{C}} \int_{z\in B_K(P,R)} |f(z) - f_{B_K(P,R)}|\, dV(z) \equiv \|f\|_* \leq C < \infty.$$

Here $f_{B_K(P,R)}$ denotes the average of the function f over the metric ball. Those familiar with such spaces of functions of bounded mean oscillation will note that the collection of balls being tested here is rather sparse. But when this space is intersected with the space of holomorphic functions on Ω then the right subspace results: in particular, there are natural relationships between this holomorphic function space and certain boundary BMO spaces.

Here are the functions of bounded mean oscillation in the boundary of Ω : If $P \in \partial\Omega$ then we let $b_1(P,r)$ be the standard Euclidean isotropic ball of radius r and center P in the boundary and we let $b_2(P,r)$ be the non-isotropic ball (with radius r in the normal direction and $\sqrt{r}$ in complex tangential directions) as discussed in [**STE2, STE3**], [**KRA1**]. These balls non-isotropic balls are the right ones for strongly pseudoconvex domains. They arise naturally in the study of Fatou type theorems and in the study of canonical kernels (see the material in Sections 1.1-1.6 and [**KRA8**]).

Now we use the standard Hausdorff area measure on $\partial\Omega$ to define two BMO spaces on $\partial\Omega$: $BMO_j(\partial\Omega)$ consists of those f on $\partial\Omega$ satisfying

$$\sup_{\substack{P\in\Omega \\ r>0}} \inf_{c\in\mathbb{C}} \int_{b_j(P,r)} |f(\zeta) - c|\, d\sigma(\zeta) < \infty.$$

Note once again that in the definition of $BMOs$ we fix the radius of the balls once and for all whereas in the definitions of the $BMO_j(\partial\Omega)$ we allow the radii to vary. This is true in part because, for the solid BMO, the complex geometry automatically takes care of the size of the balls. In particular, the completeness of the Kobayashi geometry makes the balls of fixed radius R shrink (in the Euclidean sense) as the center moves towards the boundary. The boundary, by contrast, does not have its own intrinsic complex geometry. So we must use balls of arbitrary radii.

We let $BMOAs(\Omega)$ denote the holomorphic members of $BMOs$. We let $BMOA_j(\partial\Omega)$ denote the set of those $f \in H^1(\Omega)$ (the Hardy space on Ω) with boundary limits lying in $BMO_j(\partial\Omega)$. In this last definition, the use of H^1 is primarily one of convenience. We wish to define a function f to be in the space by delimiting its boundary function $\tilde{f}$. Thus we must formulate the definition in a universe for which we know $\tilde{f}$ will exist. The space H^2, or several other choices, would have worked as well. [Lurking in the background here is the John-Nirenberg theorem.]

For the record, we record here the facts that $BMO_1 \neq BMO_2$ and there are in fact no inclusions. However $BMOA_1 \subseteq BMOA_2$ when Ω is strictly pseudoconvex, and the inclusion is strict (see [**KRA12**]). The dual of the holomorphic Hardy space $\mathcal{H}^1(\Omega)$, when Ω is strictly pseudoconvex, is $BMOA_2$ (see [**KL1**]).

In fact this last assertion does not appear explicitly in the literature but has been part of the folklore for some time. A formal proof may be obtained by imitating classical arguments. But there are subtleties involved when one is required to extend the standard boundary asymptotics ([**FEF**], [**BMS**]) for the Szegö kernel from boundary points to interior points. These matters are discussed in detail in [**KL1**] and in Section 2.3 of this book. Once those estimates are in place, then certain gradient expressions may be proved to be Carleson measures and the proof follows classical lines. From it one may obtain suitable atomic decompositions for $\mathcal{H}^1$, factorization theorems for functions in Hardy classes, duality theorems for all $H^p, 0 < p < \infty$, and so forth. See [**KL1**], [**KL2**].

There are a number of important inclusion relations among the Bloch space and the various BMOA spaces that we have introduced. Here are some of them:

$$BMOA_1 \subseteq BMOA_2 = P(L^\infty) = (\mathcal{H}^1)^* \subseteq BMOAs = \mathcal{B} = (A^1)^*.$$

Here a word of explanation is in order. Of course $\mathcal{B}$ denotes the Bloch space as defined in the beginning of this section. Also P denotes the Bergman projection. The space A^p is the Bergman space of pth power integrable functions *with respect to* dV on Ω. Finally, $\mathcal{H}^1$ is the Hardy space of holomorphic functions, uniformly integrable over level sets of $\partial\Omega$. This latter space (as well as duality questions) is considered in further detail in Section 1.3.

We refer the reader to [**BEK1**]-[**BEK3**] and [**BEBCZ**] for related results on various bounded symmetric and Siegel domains.

2.2. Normal Functions and Normal Families. The concept of normal function is a cornerstone of the function theory of one complex variable but is less familiar in the several variable setting. It is a beautiful idea, and deserves to be better known. We refer to the material in Sections 1.5, 1.6 on the Lindelöf principle for motivation.

A proper understanding of normal *function* entails first nailing down what we mean by a normal *family.*

DEFINITION 2.2.1. A family of meromorphic functions $\mathcal{F} = \{f_\alpha\}$ on a domain $U \subseteq \mathbb{C}$ is said to be *normal* if every subsequence $\{f_{\alpha_j}\}$ either has

(i) a subsequence $\{f_{\alpha_{j_k}}\}$ that converges uniformly on compact sets (i.e., that *converges normally*)

or

(ii) a subsequence $\{f_{\alpha_{j_k}}\}$ with the property that for every pair of compact sets $K \subseteq U$ and $L \subseteq \mathbb{C}$ there is a number $N = N(K, L)$ such that when $k \geq N$ then $f_{\alpha_{j_k}}(K) \cap L = \emptyset$ (i.e. that *compactly diverges*)

The point of this definition, which we learned from [**WU1**], is that we should not treat the point at infinity differently from other points of the Riemann sphere $\widehat{\mathbb{C}}$. If we think think of a meromorphic function as a holomorphic mapping from the complex manifold $D \subseteq \mathbb{C}$ to the complex manifold $\widehat{\mathbb{C}}$ (the Riemann sphere) then we may simply say that a family of meromorphic functions is normal if every subsequence has a subsequence that converges uniformly on compact sets.

It is instructive to bear in mind the simple example of the functions $\mathcal{F} = \{z^j\}, j \geq 1$, on the complex plane. If U is a subset of the disc, then $\mathcal{F}$ is normal by the first part of the definition. If U is a subset of $\{|\zeta| > 1\}$ then $\mathcal{F}$ is normal by the second part of the definition. But if U is an open set that contains a part of the circle, then $\mathcal{F}$ is not normal.

On the disc, let us use the following working definition of normal function.

DEFINITION 2.2.2. A holomorphic function $f : D \to \widehat{\mathbb{C}}$ is said to be *normal* if, whenever $\{\phi_j\}$ are conformal self-maps of the disc then $\{f \circ \phi_j\}$ form a normal family.

The paradigm for a normal function is any bounded analytic function. By Montel's theorem, such a function is plainly normal. However the reciprocal of any normal function is normal (check the definition of normal family!), and that provides us with a number of new examples of normal functions.

Even more interesting, any meromorphic function that omits three values in its range is normal. For we may assume, by composition with a linear fractional transformation, that one of the omitted values is ∞. Then we are dealing with a standard *holomorphic* function that omits two values. Post-compose with the elliptic modular function to obtain a bounded analytic function. This is plainly normal, hence so is the original function.

If f is an analytic function that omits one non-zero value in its range and if $\sqrt{f}$ makes sense then $\sqrt{f}$ must omit two (finite) values, together with ∞, and hence be normal.

As a final example, we mention that any Schlicht function on the disc is normal. This is so because the image of such a function omits a continuum on the Riemann sphere, hence certainly omits three values.

While H^∞ forms a natural subspace of the normal functions, it is the case that the normal functions have no natural relationship with H^p when $p < \infty$ (However some interplay was discovered in [**CIK**]. See also the analogous discussion of Bloch functions in Section 2.1.). Every Bloch function is normal: this is not yet plain but will become clear momentarily. In a sense, a normal function is a Bloch function that is allowed to wander off to infinity.

It is known that if f is Bloch then $\exp(f)$ is normal. In fact Flavia Colonna [**COL**] has capitalized on this fact to create a new function space. Call a holomorphic function h on the disc semi-Bloch if for every $\lambda \in \mathbb{C}$ the function $f_\lambda(\zeta) = \exp(\lambda h(\zeta))$ is normal. She proved that there exist semi-Bloch functions that are not Bloch. These ideas have not been explored in the several variable context.

The key to a clearer understanding of normal functions is Marty's theorem (see [**AHL1**]):

THEOREM 2.2.3. *Let $\mathcal{F} = \{f_\alpha\}$ be a family of (holomorphic) functions on a domain $\Omega \subseteq \mathbb{C}$. The family $\mathcal{F}$ is normal if and only if there is a constant $C > 0$ such that*

$$\frac{|f'_\alpha(\zeta)|}{1 + |f_\alpha(\zeta)|^2} \le C \frac{1}{1 - |\zeta|^2}$$

on Ω.

Recall that Montel's theorem just boils down to (i) Cauchy estimates on compact sets and (ii) the Ascoli-Arzela theorem. Marty's theorem is just the spherically invariant version of Montel's theorem. This observation will be developed in what follows. Accepting Marty's theorem, we have:

PROPOSITION 2.2.4. *A function f on the disc is normal if and only if there is a constant $C > 0$ such that*

$$\frac{|f'(\zeta)|}{1 + |f(\zeta)|^2} \leq C \frac{1}{1 - |\zeta|^2}.$$

The proposition is straightforward to verify, once one observes that f is normal if and only if $f \circ \mu$ is normal for any holomorphic function μ with range in the disc. The proposition makes some of our other assertions clear: that the reciprocal of a normal function is normal, and that a Bloch function must be normal.

Classically, the prime motivation for studying normal functions is the beautiful theorem of Lehto and Virtanen [**LEV**]:

THEOREM 2.2.5. *Let f be a normal function on the disc. Let $\gamma : [0, 1] \to \bar{D}$ be any continuous curve such that $\gamma(1) = 1 \in \partial D$. If $f \circ \gamma$ has limit ℓ at 1 then in fact f has a non-tangential limit ℓ at 1.*

The proof of this theorem is a nice exercise in hyperbolic geometry. We cannot include it here.

Marty's theorem, and the subsequent proposition, are the key to getting a handle on normal functions of several complex variables. We shall present now some ideas from [**CIK**], and refer the reader to that source for further references.

The main issue in several complex variables is that a generic domain has few, if any, biholomorphic self-maps. But one realizes after a while that, for the purposes of function theory, the symmetries represented by biholomorphic self-maps are in fact misleading. The necessary structure can instead be expressed with (hyperbolic) geometry. In fact let us take this occasion to draw out this theme using another touchstone. The theory of Möbius invariant function spaces on the ball has been an area of fruitful activity in recent years (see for instance [**RUD1**], [**RUD3**], [**AF2**] [**AFP**], [**ZHU1**], [**ZHU2**], [**ZHU3**], [**PEL**]). But it has a touch of artificiality because it cannot be meaningfully transferred to other domains (which lack large symmetry groups). It seems clear that the geometric approach, such as we are about to apply to the study of normal functions, might in fact allow a theory of "Möbius invariant-like" spaces to be developed on almost any domain.

As usual, the key to our approach is the Kobayashi metric. When studying Bloch functions, we equipped the *domain* in question with the Kobayashi metric, and we used the *Euclidean metric* on the range. What we do now is to change the metric on the range to the spherical metric. That is, we think of the range as the Riemann sphere and give that sphere the inherited metric from three dimensional Euclidean space.

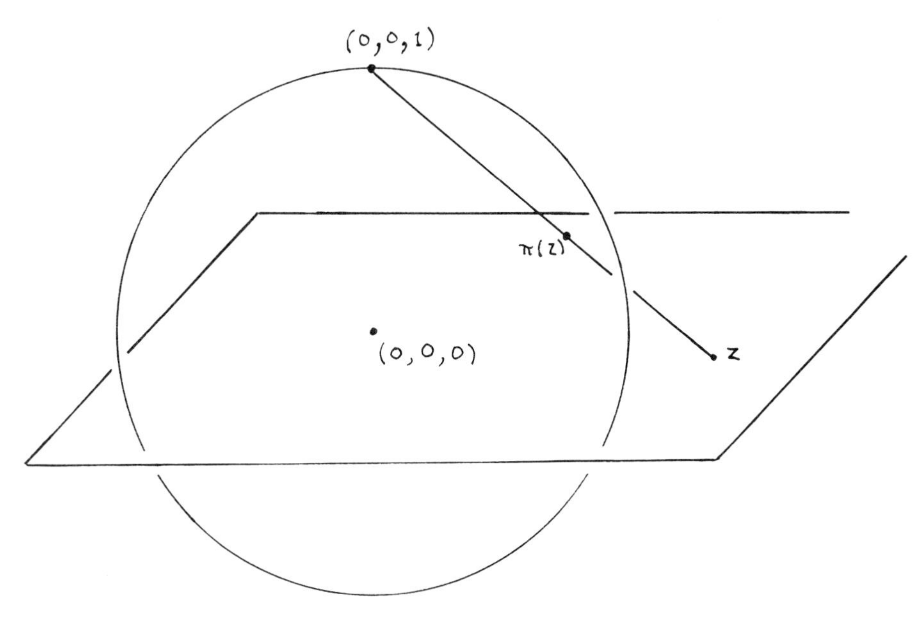

FIGURE 2.1

Let us develop the idea of the spherical metric a bit more precisely. The stereographic projection π depicted in Figure 2.1 is given in coordinates by

$$z = x + iy \to \left(\frac{2x}{|z|^2+1}, \frac{2y}{|z|^2+1}, \frac{|z|^2-1}{|z|^2+1} \right).$$

If $z = x + iy$ and $w = u + iv$ are points of the plane then their (integrated) distance in the spherical metric is just the ordinary Euclidean 3-space distance between the points $\pi(z)$ and $\pi(w)$. A calculation shows that this distance is

$$\xi(z, w) = \frac{2|z - w|}{\sqrt{1 + |z|^2}\sqrt{1 + |w|^2}}.$$

Here, and in what follows, the expression $| \quad |$ denotes Euclidean length or distance.

Now what about the infinitesimal form of the spherical metric? We know that the geodesics in the sphere are great circles. The corresponding planar geodesics are just pullbacks of these great circles under π : that is, they are lines or great circles. Now fix a point $z \in \mathbb{C}$ and consider a geodesic γ emanating from z. So $\gamma(0) = z$. For simplicity, let us suppose that γ traces out a line (although it does not really make any difference). For $\epsilon > 0$ we have that

$$\begin{aligned} \frac{2|\gamma(0) - \gamma(\epsilon)|}{\sqrt{|\gamma(0)|^2 + 1}\sqrt{|\gamma(\epsilon)|^2 + 1}} &= \chi(\gamma(0), \gamma(\epsilon)) \\ &= \int_0^\epsilon \|\gamma'(t)\|_{sph}\, dt. \end{aligned}$$

Then

$$\lim_{\epsilon \to 0^+} \frac{1}{\epsilon} \frac{2|\gamma(\epsilon) - \gamma(0)|}{\sqrt{|\gamma(0)|^2 + 1}\sqrt{|\gamma(\epsilon)|^2 + 1}} = \lim_{\epsilon \to 0^+} \int_0^\epsilon \|\gamma'(t)\|_{sph}\, dt \cdot \frac{1}{\epsilon}.$$

The right side obviously equals $\|\gamma'(0)\|_{sph}$. The left side is

$$\lim_{\epsilon \to 0^+} \left| \frac{\gamma(\epsilon) - \gamma(0)}{\epsilon} \cdot \lim_{\epsilon \to 0^+} \frac{2}{\sqrt{|\gamma(0)|^2 + 1}\sqrt{|\gamma(\epsilon)|^2 + 1}} \right| = |\gamma'(0)| \cdot \frac{2}{1 + |\gamma(0)|^2}.$$

We conclude that

$$\|\gamma'(0)\|_{sph} = \frac{2|\gamma'(0)|}{1 + |\gamma(0)|^2}.$$

[Notice that this calculation vindicates Theorem 2.2.3.]

Thinking of $\gamma(0) = z$ and $\gamma'(0) = \xi$, we are thus led to the following definition:

DEFINITION 2.2.6. Let $\mathbb{C}$ be the complex plane. The spherical metric at a point $\zeta \in \mathbb{C}$ applied to the vector $\xi \in \mathbb{C}$ is given by

$$\|\xi\|_{\mathrm{sph},\zeta} \equiv \frac{2|\xi|}{1 + |\zeta|^2}.$$

The integrated form of the metric is

$$d_S(z, w) = \frac{2|z - w|}{(1 + |z|^2)^{1/2}(1 + |w|^2)^{1/2}}.$$

Observe that this definition (and the ideas that led to it) show that Marty's theorem (2.2.3) is just a spherical version of the Ascoli-Arzela theorem.

Now we have:

DEFINITION 2.2.7. Let $\Omega \subset\subset \mathbb{C}^n$ be a domain, equipped with the Kobayashi metric. Let $\widehat{\mathbb{C}}$ be the Riemann sphere equipped with the spherical metric. A function

$$f : \Omega \to \widehat{\mathbb{C}}$$

is said to be *normal* if there is a constant $C > 0$ such that for any vector $\xi \in \mathbb{C}^n$ it holds that

$$\|f_*(z)\xi\|_{sph} \leq C\|\xi\|_{\mathrm{Kob}}.$$

Observe that the tangent vector ξ did not come up in our Marty-like characterization of normal functions in one complex variable because the tangent space in that context is one dimensional, hence the vector can be divided out of the inequalities. Now the presence of the tangent vector is essential: in particular, normal directions near the boundary will be treated differently from tangential directions.

We should mention right away that, in the function theory of several complex variables, it is somewhat artificial to consider functions taking values in $\widehat{\mathbb{C}}$. In fact an analogous theory for functions valid in any complete Hermitian manifold has been developed in [**ALA2**].

The reader is invited to check that, since the Kobayashi metric on the disc has size $1/(1-|\zeta|^2)$, the definition we have just given is consistent with the classical definitions of normal function on the disc given above.

The following result from [**CIK**] validates that we have the right definition of normal function:

THEOREM 2.2.8. *Let $\Omega \subseteq \mathbb{C}^n$ be a bounded domain with C^2 boundary. Let $f : \Omega \to \widehat{\mathbb{C}}$ be a holomorphic function (the target space may be replaced by a complex manifold, which for many purposes is more natural—see* [**ALA2**]*). Then the following are equivalent:*

1: *f is normal;*

2: *For every sequence $\phi_j : D \to \Omega$ it is the case that $\{f \circ \phi_j\}$ is a normal family;*

3: *For every domain $\Omega' \subseteq \mathbb{C}^{n'}$ and holomorphic mapping $\Phi : \Omega' \to \Omega$ it holds that $f \circ \Phi$ is normal;*

4: *For any $\mathbf{k}$ a complex affine subspace of $\mathbb{C}^n$ having non-trivial intersection with Ω it holds that $f\big|_{\mathbf{k}\cap\Omega}$ is normal.*

We omit the details of the proof, which are not particularly interesting. But it is clear by now that we have the right definition of normal function. Incidentally, on the ball B (the only smoothly bounded domain with transitive automorphism group) a function is normal if and only if, whenever $\{\Phi_j\}$ are biholomorphic self-maps of B then $\{f \circ \Phi_j\}$ is a normal family.

The final vindication of the given definition of normal function is that these functions satisfy a Lindelöf principle. Recall from Section 1.5 that the Lindelöf principle *does not work* on the ball in $\mathbb{C}^2$, even for bounded analytic functions, if one uses the admissible approach regions of Koranyi. Thus a more restrictive method of approach is required. This is the idea of "hypoadmissible convergence."

DEFINITION 2.2.9. Let $\Omega \subseteq \mathbb{C}^n$ be a bounded domain with C^2 boundary. Let $\mathcal{H} \subseteq \Omega$ be an open, connected subregion. We call $\mathcal{H}$ a *hypoadmissible* approach region at the boundary point P if the following properties hold:

1: $\mathcal{H} \supseteq \Gamma_\alpha(P)$ for some non-tangential approach region Γ_α at P;

2: $\mathcal{H} \subseteq \mathcal{U}_\alpha(P)$ for some "Kobayashi approach region $\mathcal{U}_\alpha(P)$" as defined in Sections 1.5, 1.6;

3: It holds that

$$\lim_{\epsilon \to 0^+} \sup_{z \in B(P,\epsilon)\cap\mathcal{H}} \operatorname{dist}\big(z\,,\, B(P,\epsilon)\cap\Gamma_\alpha(P)\big) = 0.$$

We say that a function f on Ω has *hypoadmissible limit* ℓ at $P \in \partial\Omega$ if it has the limit ℓ through every hypoadmissible approach region.

So, on the ball for instance, a hypoadmissible approach region is one that is asymptotically smaller (in complex tangential directions) than an admissible approach region of Koranyi (recall that, on the ball, admissible approach regions and Kobayashi approach regions are comparable). On the disc, hypoadmissible approach regions and non-tangential approach regions are just the same. Now a typical theorem is this:

THEOREM 2.2.10. *Let $\Omega \subseteq \mathbb{C}^n$ be a bounded domain with C^2 boundary. Let f be a normal function on Ω. Let $P \in \partial\Omega$. If f has radial limit ℓ at P then f has hypoadmissible limit ℓ at P.*

The proof of this theorem follows from simple metric inequalities and we omit the details.

Unfortunately, a theorem such as Lehto and Virtanen's—which allows the curve in the hypothesis to lie in the boundary and concludes that the same limit exists in the non-tangential sense—is not true in several complex variables. Consider the example from Section 1.4: $f(z_1, z_2) = z_2^2/(1 - z_1)$. The curve $\gamma(t) = (t, \sqrt{1-t^2})$ lies in ∂B (in fact it is complex tangential—see Section 1.1). And $\lim_{t\to 1^-} f \circ \gamma(t) = 2$. However the hypoadmissible limit of f at the terminal point $(1, 0)$ is 0. To further complicate matters, some boundary curves cause no trouble. For example, the curve $\tilde{\gamma}(t) = (\sqrt{2t - t^2}, 1 - t)$ lies in the boundary (it is complex normal), and the limit of $f \circ \gamma$ as $t \to 1$ is 0. This agrees with the hypoadmissible limit at $(1, 0)$—not only for this particular f but (as one can prove) for any normal function f.

The problem of course—that is the yardstick for telling good boundary curves from bad ones—is whether the curve is complex tangential or complex normal. In fact some small Tauberian hypothesis is required to get a theorem. First we introduce a standard bit of notation. If $P \in \partial\Omega$ and f is a function on Ω then we let $|f(P)| = \limsup_{\Omega \ni z \to P} |f(z)|$. [This notation is a bit nonstandard in mathematical analysis, but it is accepted in this subject—see [**LEV**].] Here is the best known result:

THEOREM 2.2.11. *Let $\Omega \subseteq \mathbb{C}^n$ be a bounded domain with C^2 boundary. Let $\gamma : [0, 1] \to \partial\Omega$ be a C^2 curve which is complex normal (that is, $\dot{\gamma}(t)$ is not in the complex tangent space for any t. Let $f : \Omega \to \widehat{\mathbb{C}}$ be holomorphic. If either*
(1) *$f \in H^p(\Omega)$ for some $p > 4n$*
or
(2) *$f \in H^p(\Omega)$ for some $p > 0$ and the image of f omits two values*
then the following holds: if $\lim_{t\to 1^-} |f \circ \gamma(t)| = 0$ then f has hypoadmissible limit 0 at $P = \gamma(1)$.

The proof of this theorem is hard work. The idea is that the curve γ bounds an almost analytic disc in Ω. One applies the theorem of Lehto and Virtanen to the function f restricted in this disc to obtain a nontangential curve in Ω that terminates at P and along which f has the required limit. Then we can apply the Lindelöf principle of several variables to finish the proof.

We cease our discussion of Lindelöf principles and normal functions at this point. It is my view that the subject is not in a satisfactory state. The dialectic

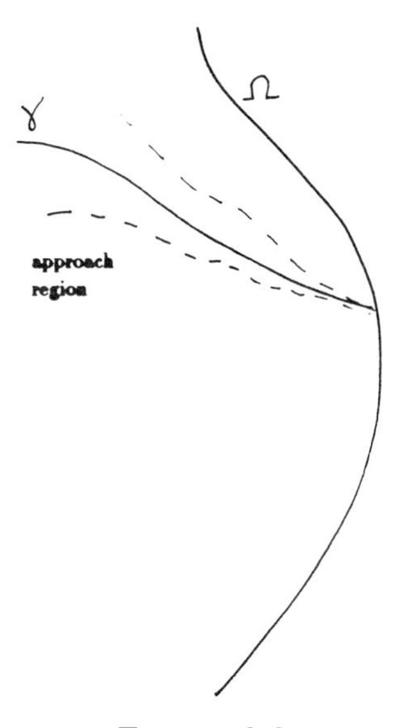

FIGURE 2.2

between complex tangential curves and complex normal curves is not completely understood. Hypoadmissible approach regions are too small to be completely satisfactory. If the approach curve γ in the hypothesis of the Lindelöf principle is too near to being a complex tangential curve in the boundary—in fact even if it runs along the parabolic edge of an admissible approach region—then a bounded or normal function will not generally have the same hypoadmissible limit at the terminal point. There *will* be a region through which the function does have the right limit, but the region will be one defined using the Kobayashi distance *not* to the inward normal but to the curve γ in the hypothesis. The picture is as in Figure 2.2. This area deserves to be better understood.

We would like to close the section with a brief treatment of a criterion for normality that was inspired by a lecture of Abraham Robinson (see [**ROB**]). He enunciated the heuristic that "any property that would tend to make an entire function constant would also tend to make a family of functions normal." An example of such a property is *boundedness.* Larry Zalcman [**ZAL1, ZAL2**] found a concrete enunciation of Robinson's principle. However his formulation relied heavily on the translation structure of the complex plane and its obvious generalization to several complex variables turned out to be false (see [**HAH2**]). We seek here to find a version of Robinson's principle that is valid in some generality. Here is one effort. We let the symbol $\mathcal{P}$ denote a property in the sense of Robinson. By this we mean a collection of functional elements $\langle f, \Omega \rangle$, where f is a holomorphic function on the open set $\Omega \subseteq \mathbb{C}^n$. Let Δ_r denote the disc in the complex plane with center 0 and radius r.

THEOREM 2.2.12 (ALADRO/KRANTZ [**ALAK**]). *Let $\mathcal{P}$ be a property of holomorphic function which satisfies the following conditions:*

1: *If* $\langle f, \Omega\rangle \in \mathcal{P}$ *and* $\Omega' \subset \Omega$ *then* $\langle f, \Omega'\rangle \in \mathcal{P}$.

2: *Let us define*

$$\mathcal{P}_1 = \{\langle f \circ \phi, \Delta_r\rangle : \langle f, \Omega\rangle \in \mathcal{P} \text{ and } \phi \in Hol(\Delta_r, \Omega) \text{ for some } r > 0\}.$$

Let $\{r_j\}$ *be a sequence satisfying* $0 < r_1 < r_2 < \cdots$ *and* $r_j \to +\infty$. *Suppose that* $\phi_j \in Hol(\Delta_{r_j}, \Omega), (f_j, \Omega) \in \mathcal{P}$, *and that* $\langle f_j \circ \phi_j, \Delta_{r_j}\rangle \in \mathcal{P}_1$. *If* $f_j \circ \phi_j \to g$ *uniformly on compact subsets of* $\mathbb{C}$, *then* $\langle g, \mathbb{C}\rangle \in \mathcal{P}_1$.

3: *If* $\langle g, \mathbb{C}\rangle \in \mathcal{P}_1$, *then* g *is a constant function.*

Then, for any domain $\Omega \subseteq \mathbb{C}^n$, *the family of functions satisfying* $\langle f, \Omega\rangle \in \mathcal{P}$ *is normal on* Ω.

The reader familiar with classical complex function theory will recognize here elements of the theorem of Lohwater/Pommerenke [**LOP**] on normal functions and the characterization by Zalcman [**ZAL2**] of non-normal families. Also present are the sort of "cone of functions" conditions that are used to describe the Hartogs functions on a domain (see [**KRA1**]).

2.3. H^1, BMO, and $T(1)$.

Introduction

It has been "well known" for many years that the dual of H^1 is BMO under rather general circumstances. In particular, with the atomic definition of (real variable) H^1, it is virtually automatic that it dual is H^1 (see Section 1.3). What has been swept under the rug is the characterization of the dual of the *holomorphic* Hardy space, which in this section we shall refer to as $\mathcal{H}^1$. That the dual of $\mathcal{H}^1$ is $BMOA$ (suitably defined) when the domain is the unit ball in $\mathbb{C}^n$ appeared in an unpublished version of [**CRW**]. The arguments related to these matters that appear in [**GAL**] are incomplete.

The version of the Riesz transforms that appear in [**CHG**]—a powerful tool that merits further exploration—could also be used to study the duality issue. But all of the papers mentioned thus far use special symmetries of the ball. In particular, they use known explicit formulas for the Bergman and Szegö kernels when the domain is the ball.

Actually carrying out the proof in the more general context of strongly pseudoconvex domains involves a great deal of technical machinery; for many of the needed estimates on the Szegö kernel, for instance, the standard asymptotic expansions (see [**FEF**] and [**BOS**]) are not quite of the form required. More precisely, the asymptotic formulas are stated for $S(z, \zeta)$ *when both variables lie in the boundary.* In the proofs, one needs estimates for $z \in \Omega$ and $\zeta \in \partial\Omega$.

In the present section we shall sketch how to establish that $(\mathcal{H}^1)^*(\Omega) = BMOA(\Omega)$ when either Ω is strongly pseudoconvex or Ω is of finite type in $\mathbb{C}^2$. What distinguishes these two types of domains is that we have detailed estimates on the canonical kernels in those situations. Along the way, we shall note how the $T(1)$ theorem of David/Journé can be used to shed some light on these issues. At the end, we make some comments on atomic decompositions.

In effect, this section is an introduction to the paper [**KL1**]. We note that Galia Dafni [**DAF**] has developed some of these ideas using entirely different methods—namely those connected with the Calderón reproducing formula and its variants. The Calderón formula that she has developed will prove to be a powerful tool in future studies.

The Concept of the $T(1)$ Theorem

While the $T(1)$ theorem was originally proved in order to put the study of variable coefficient singular integral operators (especially their L^2 theory) into a more complete form, it was established in [**NRSW**] that the philosophy of the $T(1)$ theorem is a powerful ally in the harmonic analysis—especially the L^p theory, $1 < p < \infty$, of domains in $\mathbb{C}^n$. More recent results of [**HSA1**], [**HSA2**], in which the conclusions of the $T(1)$ theorem apply to the Triebel-Lizorkin spaces on a space of homogeneous type, extend the utility of this circle of ideas. In particular, the $T(1)$ theorem, suitably formulated, is valid on Triebel-Lizorkin spaces in the context of a space of homogeneous type.

In fact, to be true to the literature, we should make a disclaimer at this point. Technically speaking, the original $T(1)$ theorem was a device for verifying the L^2 boundedness of certain variable coefficient integral operators. In the translation invariant case, Fourier analysis and other classical devices could serve instead. But for a kernel $K(x, y)$ this was a difficult issue and the David/Journé theory cut to the heart of the matter. The book [**CHR1**] is careful to distinguish this theorem from generalities evolving from the Calderón/Zygmund theory that give boundedness, for instance, on other L^p spaces. More recent work has tended, at least in this author's mind, to blur this distinction. Moreover, the papers [**NRSW**], [**FTW**], [**HJMT**], [**HSA1**], [**HSA2**] [**FRJ**] (to name a few) have further erased the distinction by viewing the $T(1)$ theorem and its hypotheses as a device for seeing that an operator is bounded on all of the Triebel-Lizorkin spaces. At the risk of exasperating some readers, I shall view "singular integral" theory and "$T(1)$ theory" as part of the same gestalt, not taking great care to distinguish the two.

We begin by briefly stating the $T(1)$ theorem in its classical setting. This will facilitate understanding of the result in the more general context that we shall study below. The class of operators that we shall study is the *generalized Calderón-Zygmund operators.* Such an operator $T : \mathcal{D} \to \mathcal{D}'$ is induced (by the classical theorem of L. Schwartz) by an integral kernel K that satisfies the following conditions:

1: $K \in \mathcal{D}'(\mathbb{R}^N \times \mathbb{R}^N)$
2: If $\phi, \psi \in \mathcal{D}$ then

$$\langle T\phi, \psi\rangle = \langle K, \psi \otimes \phi\rangle;$$

3: Off the diagonal, K is a continuous function of its two variables;

4: There is a constant $C > 0$ such that

$$|K(x,y)| \le C \cdot |x-y|^{-N};$$

5: There are constants $C, \epsilon > 0$ such that whenever $2|y-y'| \le |x-y|$ then

$$|K(x,y) - K(x,y')| + |K(y,x) - K(y',x)| \le C|y-y'|^{\epsilon}|x-y|^{-N-\epsilon}.$$

The first three properties listed here are plainly technical ones; the main point is the last two. Following [**FTW**], it will be convenient for us to refer to an operator satisfying **1** - **5** as an element of CZO(ϵ).

David and Journé [**DJS**] proved that necessary and sufficient for a generalized Calderón-Zygmund operator to be bounded on $L^2(\mathbb{R}^N)$ is that the following three properties are satisfied:

A: Let 1 denote the constant function that is identically equal to 1. Then $T(1)$, suitably interpreted, lies in BMO.

B: Let T^* denote the adjoint of T. Then T^*1 lies in BMO.

C: The operator T satisfies the *weak boundedness property*: for every bounded subset $\mathcal{B} \subseteq \mathcal{D}$ there exists a positive constant $C = C(\mathcal{B})$ such that for all $\phi, \psi \in \mathcal{B}$, all $x \in \mathbb{R}^N$, and all $t > 0$ it holds that

$$\left|\langle T\phi^{x,t}, \psi^{x,t}\rangle\right| \le Ct^N.$$

Here the function $\phi^{x,t}(y)$ is defined to equal $\phi((y-x)/t)$ and $\psi^{x,t}$ similarly.

Of course the use of the constant function 1 in criteria **A** and **B** is a matter of convenience; there is a dense set of functions in L^∞ that will do as well. In the classical setup, weak boundedness is virtually automatic by scaling.

A recent development in the theory of singular integrals (see [**HSA1**], [**HSA2**]) is that they are bounded on certain Triebel-Lizorkin spaces. These spaces include Lebesgue, Lipschitz, Sobolev, Besov, Nikolski'i, Hardy, and other classical spaces. We describe them briefly in the next subsection.

The Triebel-Lizorkin Spaces

For the sake of efficacy, we merely summarize one of the original definitions of the Triebel-Lizorkin spaces (coming from Littlewood-Paley theory) and then indicate the perhaps more useful characterization due to Frazier and Jawerth [**FRJ**]. Fix $\phi \in \mathcal{S}$, the Schwartz space, such that

a: $\operatorname{supp}\hat{\phi} \subseteq \{\xi \in \mathbb{R}^N : 1/2 \le |\xi| \le 2\}$;

b: $|\hat{\phi}(\xi)| \ge \epsilon > 0$ for all $3/5 \le |\xi| \le 5/3$;

c: For ν a positive integer we set $\phi_\nu(x) = 2^{\nu N}\phi(2^\nu x)$. Then the identity

$$\sum_{\nu \in \mathbb{Z}} |\hat{\phi}_\nu(\xi)|^2 \equiv 1$$

holds for all $\xi \neq 0$.

DEFINITION 2.3.1. Let $\alpha \in \mathbb{R}$ and $1 \leq p, q \leq \infty$. Let $\mathcal{P}$ be the space of all polynomials on $\mathbb{R}^N$. Let $f \in \mathcal{S}'(\mathbb{R}^N)/\mathcal{P}$. We say that $f \in \dot{F}_p^{\alpha,q}$ if

$$\|f\|_{\dot{F}_p^{\alpha,q}} \equiv \left\| \left(\sum_{\nu \in \mathbb{Z}} [2^{\nu\alpha} |\phi_\nu * f|]^q \right)^{1/q} \right\|_{L^p} < \infty.$$

It turns out that this definition is independent of the choice of ϕ (that is, the norm may change in size but the space remains the same). For concreteness, we point out that

$$\begin{aligned} \dot{F}_p^{0,2} &\approx L^p \,, 1 < p < \infty \\ \dot{F}_p^{0,2} &\approx H^p \,, 0 < p \leq 1 \\ \dot{F}_\infty^{0,2} &\approx BMO \end{aligned}$$

For $\alpha > 0$, $\dot{F}_\infty^{\alpha,\infty}$ is (the homogeneous version of) the classical Lipschitz space of order α and for $\alpha > 0, 1 \leq p < \infty, 1 \leq q \leq \infty$ it holds that $\dot{F}_\alpha^{p,q}$ is a Sobolev space.

An appealing characterization of Triebel-Lizorkin spaces using decompositions into "smooth atoms" or "smooth molecules" has been obtained by Frazier and Jawerth ([**FRJ**] and references therein). We shall say no more about them here, but refer the reader to [**FTW**] for background on these matters. It is these two decompositions that provide the key to proving the versions of the $T(1)$ theorem that are presented below.

We now present a version of the $T(1)$ theorem for Triebel-Lizorkin spaces (see [**FTW**]). Technical notation in this theorem is as follows: if β is a real number then $[\beta]$ is the greatest integer less than or equal to β and $(\beta) \equiv \beta - [\beta]$.

THEOREM 2.3.2. *Let $\alpha \geq 0$ and min $\{p, q\} \geq 1$. Assume that the linear operator $T : \mathcal{D} \to \mathcal{D}'$ satisfies*

2.3.2.1: *$T \in CZO([\alpha] + \delta)$, some $(\alpha) < \delta < 1$;*
2.3.2.2: *$T^* \in CZO(\rho)$ for some $0 < \rho < 1$;*
2.3.2.3: *T satisfies the weak boundedness property;*
2.3.2.4: *$T(y^\gamma) \in BMO$ for all $|\gamma| \leq [\alpha]$;*
2.3.2.5: *In case $\alpha = 0$ then $T^*1 \in BMO$.*

Then T extends to be a bounded operator on $\dot{F}_p^{\alpha,q}$.

It requires some extra work to see how, in property 2.3.2.4, T is extended to act on polynomials (this is in fact done by a limiting argument with cutoff functions—see [**FTW**] for details). The special case of the theorem in which we are interested is $\alpha = 0$. Then condition 2.3.2.1 becomes $T \in CZO(\delta)$ for some $0 < \delta < 1$ and condition 2.3.2.4 becomes $T(1) \in BMO$ as in the classical case. We note in passing that it is a standard fact (again see [**FTW**]) that proving the theorem under the conditions $T(1) \in BMO$ and $T^*(1) \in BMO$ can be reduced to the hypothesis that $T(1) = 0$ and $T^*(1) = 0$. See [**FTW**] for details.

A Version of the $T(1)$ Theorem on Spaces of Homogeneous Type

Refer to the material in Chapter 1, to Section 2.6, and to [**COW1**], [**COW2**] for the concept of "space (X, d, μ) of homogeneous type." A useful tool in the study of spaces of homogeneous type is the notion of "measure distance." The measure distance ρ on X is defined for $x, y \in X$ by

$$\rho(x,y) = \min\Big(\inf\{\mu(B(y,r)) : x \in B(y,r)\}, \inf\{\mu(B(x,r)) : y \in B(x,r)\}\Big),$$

Notice that, on Euclidean space $\mathbb{R}^N$ equipped with Lebesgue measure, the measure distance between two points is comparable to the Nth power of the usual Euclidean distance.

While spaces of homogeneous type retain many of the favorable features of Euclidean space, they have in general no translation structure and no differentiable structure. We describe now a substitute framework for treating generalized Calderón-Zygmund operators and the $T(1)$ theorem. We will say that an operator T, given by a kernel $K(x, y)$ that is continuous off the diagonal, is a generalized Calderón-Zygmund operator of order ϵ ($T \in CZO(\epsilon)$) if there are constants $c, C > 0$ such that

1: For $x, y \in X$ we have

$$|K(x,y)| \leq C \cdot \rho(x,y)^{-1};$$

2: If $x, y, y' \in X$ satisfy $\rho(y, y') \leq c\rho(x, y)$ then

$$|K(x,y) - K(x,y')| + |K(y,x) - K(y',x)| \leq C\rho(y,y')^\epsilon \rho(x,y)^{-(1+\epsilon)}.$$

Now we define the weak boundedness property on a space of homogeneous type. Our presentation comes from [**CHR1**], which we recommend for an elegant exposition of related ideas. Let $\delta \in (0, 1], x \in X$, and $r > 0$. We define the space $A(\delta, x, r)$ to be the set of $\phi \in \Lambda_\delta$ (the classical space of Lipschitz functions on a quasimetric space, defined in the obvious way) which are supported in the ball $B(x, r)$, are bounded in sup norm by 1, and satisfy

$$|\phi(y) - \phi(z)| \leq r^{-\delta} d(y,z)^\delta.$$

Note that the factor of $r^{-\delta}$ on the right is included for reasons of scaling.

DEFINITION 2.3.3. We say that a singular integral operator T on a space of homogeneous type is *weakly bounded* if there exists a $\delta > 0$ and $C < \infty$ such that for all $x \in X, r > 0$, and $\phi, \psi \in A(\delta, x, r)$ we have the inequality

$$|\langle T\phi, \psi\rangle| \leq C\mu\big(B(x,r)\big).$$

The following theorem is now a special case of the material that appears in [**HJMT**], [**HSA1**], [**HSA2**]. See also [**FTW**], Theorem 3.1.

THEOREM 2.3.4. *Let (X, d, μ) be a space of homogeneous type. Suppose that T is a generalized Calderón-Zygmund operator of some positive order ϵ.*

If $T1 \in BMO, T^*1 \in BMO$, and if T satisfies the weak boundedness property then T is bounded on all L^p and real variable H^p spaces; moreover, T is a bounded operator both on BMO and on the (atomic) Hardy space H^1.

Remark: In our applications, the space of homogeneous type will be the boundary of a smoothly bounded domain in $\mathbb{C}^n$. Therefore the space is compact. As a result the constant function 1 is in L^2 and $T1, T^*1$ are defined without difficulty.

Strongly Pseudoconvex Domains

Let $\Omega \subseteq \mathbb{C}^n$ be a strongly pseudoconvex domain with C^2 boundary. Let $\beta(\zeta, r)$ be the usual non-isotropic balls in $\partial\Omega$ (see Sections 1.2 and 1.3). Further let d be the corresponding non-isotropic quasidistance. Let $d\sigma$ be $(2n-1)$- dimensional Euclidean area measure on $\partial\Omega$. Details of these matters may be found in [**STE3**] and [**KRA1**].Then $(\partial\Omega, d, \sigma)$ is a space of homogeneous type.

Asymptotic expansions for the Szegö kernel on such a domain may be found in [**FEF**] or, more explicitly, in [**BOS**]. For our purposes in the present work, a more uniform treatment will result by noticing that the methods of the paper [**NRSW**] apply without change to strongly pseudoconvex domains in *any dimension*. They yield estimates of the following sort for the Szegö kernel S :

a: If $z, w \in \partial\Omega, X_z$ represents a complex tangential derivative at z, and X_w represents a complex tangential derivative at w, then

$$\left|X_z^I X_w^J S(z,w)\right| \le C \cdot \frac{d(z,w)^{-|I|-|J|}}{\rho(z,w)}.$$

Here I, J are multi-indices, d is the quasi-distance associated with $\partial\Omega$ when thought of as a space of homogeneous type, and ρ is the measure distance.

b: For each non-negative integer ℓ there is a constant M such that if ϕ is a C_c^∞ function supported in $B(z, \delta)$ then, with notation as in (**a**),

$$|X_z^I T\phi(z)| \le C \cdot \delta^{-\ell} \sup \sum_{|J| \le M} \delta^{|J|} |X^J \phi|$$

for all multi-indices $|I| \le \ell$.

Now let us look at the $T(1)$ theorem for a space of homogeneous type. The required growth estimates on the Szegö kernel S follow from property (**a**) just listed. The weak boundedness property follows directly from (**b**). Of course, for the Szegö operator S we have $S1 = S^*1 \equiv 1$, which is certainly in BMO. However, notice that checking the smoothness property **2.** of the kernel is a bit tricky. We shall say more about this last property below.

If we accept these properties, then we may conclude from the $T(1)$ theorem that the Szegö operator on a strongly pseudoconvex domain maps $BMO(\partial\Omega)$ to $BMO(\partial\Omega)$ and $H^1(\partial\Omega)$ to $H^1(\partial\Omega)$.

Remark: There are two potential definitions of $BMOA$, the space of holomorphic functions of bounded mean oscillation. Restrict attention to the domain $\Omega = B$, the unit ball. For convenience consider $BMOA(\partial B)$ to be a subspace of $H^2(\partial B)$, so that elements of $BMOA$ have well-defined boundary functions. Then one could model $BMOA$ either on the Euclidean geometry of isotropic balls or on the non-Euclidean geometry of non-isotropic balls. Since the Lipschitz spaces of holomorphic functions modeled on these two geometries turn out to be the same (see Section 1.1), and since BMO is in a natural sense a limit of the Lipschitz spaces Λ_α as $\alpha \to 0^+$, one might suppose that the two different $BMOA$ spaces are the same. But they are not (see [**KRA12**]). In fact the $BMOA_1$ space modeled on the non-isotropic geometry is a proper superspace of that for the Euclidean $BMOA_2$ space. (This was first proved in [**KRA12**] by an extremely indirect argument. Ullrich [**ULL**] has actually produced an explicit element of $BMOA_2 \setminus BMOA_1$.) The non-isotropic $BMOA$ space is the one that is appropriate for complex analysis of several variables, and is the one which arises in the present paper (for instance as the image of L^∞ on the boundary of the unit ball under the Szegö projection). See the additional discussion in Section 2.1. When we refer to $BMOA$ in the rest of this section (without a subscript), we shall always mean $BMOA_2$.

Domains of Finite Type in $\mathbb{C}^2$

In fact the setup that we have described for strongly pseudoconvex domains in the last subsection is also known to hold on domains of finite type in $\mathbb{C}^2$. The required estimates on the Szegö kernel are given explicitly in [**NRSW**], p. 133. The weak boundedness property is verified in detail in [**NRSW**], p. 143. The smoothness property **2.** of the kernel requires some additional arguments (see below). As a result, we may conclude that the Szegö operator on a domain of finite type in $\mathbb{C}^2$ maps $BMO(\partial\Omega)$ to $BMO(\partial\Omega)$ and $H^1(\partial\Omega)$ to $H^1(\partial\Omega)$.

The Duality of H^1 and BMO

There are two main points to be addressed in this subsection: 1) to see that the dual of the real variable Hardy space, on the boundary of either a strongly pseudoconvex domain or a domain of finite type in $\mathbb{C}^2$, is the space of functions of bounded mean oscillation; 2) to see that the Szegö projection of the real variable Hardy space H^1 is in fact the holomorphic space $\mathcal{H}^1$ from complex function theory (see [**KRA1**]). It is again part of the folklore that these facts follow from the mapping properties of the Szegö operator outlined in the last two sections. However there are some unexpected complications. We wish to sort out what comes for free and what requires additional arguments.

First let us recall how the atomic Hardy space is defined on a space (X, d, μ) of homogeneous type. We call a μ- measurable function a an *atom* if

1: a is supported in a ball $B(x, r)$;

2: The function a satisfies

$$|a(x)| \leq \frac{1}{\mu(B(x,r))}.$$

3: The function a satisfies the mean value condition

$$\int a(x)\, d\mu(x) = 0.$$

A function f on X is said to be in H^1 if there is a sequence $\{\lambda_j\} \in \ell^1$ and atoms $\{a_j\}$ such that $f = \sum_j \lambda_j a_j$. The H^1 norm of f is defined to be the infimum, over all such decompositions, of $\|\{\lambda_j\}\|_{\ell^1}$.

PROPOSITION 2.3.5. *Let (X, d, μ) be any space of homogeneous type. Let g lie in the dual space of H^1. Then $g \in BMO$.*

PROOF. By a limiting argument using the dense subspace $L^2 \cap H^1$, it is not difficult to see that the dual of H^1 consists only of locally integrable functions and that (for an *a priori* inequality) the pairing is the usual L^2 pairing.

Let g be such a function. If $B(x, r)$ is any ball and $h \in L^\infty(B(x, r))$ has essential sup norm 1 then let

$$c = \frac{1}{\mu(B(x,r))} \int_{B(x,r)} h(x)\, d\mu(x).$$

Observe that the function

$$a(x) = \frac{1}{2\mu(B(x,r))} \Big[h(x) - c \Big]$$

is an atom. Thus

$$\left| \int g(x) a(x)\, d\mu(x) \right| \leq C,$$

where the constant C is independent of the ball B and the L^∞ function h. Because a has mean value zero, we may rewrite this as

$$\left| \int [g(x) - g_B] a(x)\, d\mu(x) \right| \leq C,$$

or

$$\frac{1}{\mu(B)} \left| \int [g(x) - g_B] h(x)\, d\mu(x) \right| \leq C.$$

Here f_B denotes the mean value of f over B. But since h was an arbitrary L^∞ function of norm 1, we conclude from the last line that

$$\frac{1}{\mu(B)} \left| \int |g(x) - g_B|\, d\mu(x) \right| \leq C.$$

In other words, $g \in BMO$. ∎

PROPOSITION 2.3.6. *Let (X, d, μ) be any space of homogeneous type. If $g \in BMO$ then g lies in the dual of H^1.*

PROOF. It is enough to check that g pairs with an atom, with a bound that is independent of the particular choice of atom. If a is such an atom, with support in $B = B(x, r)$, then

$$\begin{aligned}
\left| \int g(x)a(x)\,dx \right| &= \left| \int [g(x) - g_B]a(x)\,dx \right| \\
&\leq \sup_B |a| \int_B |g(x) - g_B|\,dx \\
&\leq \frac{1}{\mu(B)} \int_B |g(x) - g_B|\,dx \\
&\leq \|g\|_{BMO}.
\end{aligned}$$

That prove the result. ∎

Now let us consider the dual of the Hardy space of holomorphic functions $\mathcal{H}^1(\Omega)$, where Ω is either a strongly pseudoconvex domain in $\mathbb{C}^n$ or a domain of finite type in $\mathbb{C}^2$. If S is the Szegö projection and H^1 the real Hardy space (for more on real Hardy spaces, see Section 2.5) then clearly $S : H^1 \to H^1 \cap \{\text{holomorphic functions} \subseteq \mathcal{H}^1\}$. Thus $S : H^1 \to \mathcal{H}^1$. In particular, if $f = \sum_j \lambda_j a_j$ is the atomic decomposition of an element $f \in H^1$ then $Sf = \sum \lambda_j Sa_j$ is a decomposition (not necessarily atomic, but in fact a *molecular decomposition*—for which see [**TAW**]) of $Sf \in \mathcal{H}^1$. We wish to see that the converse is true as well.

As a first step we have:

THEOREM 2.3.7. *The dual of $\mathcal{H}^1$ is the Szegö projection of L^∞.*

PROOF. If $g = S\gamma$ with $\gamma \in L^\infty$ then, for all $f \in \mathcal{H}^1$ we have that

$$\begin{aligned}
\langle f, g \rangle &= \langle f, S\gamma \rangle \\
&= \langle Sf, \gamma \rangle \\
&= \langle f, \gamma \rangle \\
&\leq \|f\|_{\mathcal{H}^1} \cdot \|\gamma\|_{L^\infty}.
\end{aligned}$$

Conversely, if $\alpha \in [\mathcal{H}^1]^*$ then, by the Hahn-Banach theorem, α continues to an element of $(L^1)^*$. Thus there is a function $g \in L^\infty(\partial\Omega)$ that represents this extended functional. So, for all $f \in \mathcal{H}^1$,

$$\begin{aligned}
\langle f, \alpha \rangle &= \langle f, g \rangle \\
&= \langle Sf, g \rangle \\
&= \langle f, Sg \rangle.
\end{aligned}$$

Thus the functional α is represented by Sg with $g \in L^\infty$. ∎

COROLLARY 2.3.8. *The space $S(L^\infty)$ is dense in $BMOA$.*

PROOF. The mapping $S : L^\infty \to BMOA \equiv S(BMO)$ is adjoint to the univalent mapping $i : S(H^1) \to L^1$. Since the latter is injective, the desired result follows. ∎

If one could prove that $S(L^\infty) = BMOA$ then the duality of $\mathcal{H}^1$ and $BMOA$ would follow, as would the "holomorphic atomic decomposition" for elements of $\mathcal{H}^1$. One can show, by arguments in the spirit of the last corollary, that the space of all functions of the form $\sum_j \alpha_j S a_j$, with $\{\alpha_j\} \in \ell^1$ and each a_j an atom, is *dense* in $\mathcal{H}^1$. But showing that the spaces are equal begs the duality issue.

Suppose instead that we have already established that $(\mathcal{H}^1)^* = BMOA$. Then we immediately obtain the following:

THEOREM 2.3.9. *Assume that $(\mathcal{H}^1)^* = BMOA$. It follows that an L^1 function f on $\partial\Omega$ lies in $\mathcal{H}^1$ if and only if*

$$f = \sum_j \alpha_j S a_j,$$

with $\{\alpha_j\} \in \ell^1$, each a_j an atom, and S the Szegö projection.

PROOF. If a is a 1-atom then $\|Sa\|_{\mathcal{H}^1} \leq C$. Here we use the corollary of the $T(1)$ theorem to the effect that $S : H^1_{\mathrm{Re}} \to H^1_{\mathrm{Re}}$. In particular, $Sa \in \mathcal{H}^1$, with a uniform bound on its norm. It follows that any sum of the form $\sum_j \alpha_j S a_j$ lies in $\mathcal{H}^1$.

Conversely, let $f \in \mathcal{H}^1$. We shall show that $f \in H^1_{\mathrm{Re}}$. Since we are assuming the duality of $\mathcal{H}^1$ and $BMOA$, we may proceed as follows: let $g \in BMO$ satisfy $\|g\|_* \leq 1$. Then

$$\begin{aligned} |\langle f, g\rangle| &= |\langle Sf, g\rangle| \\ &= |\langle f, Sg\rangle| \\ &\leq \|f\|_{\mathcal{H}^1} \cdot \|Sg\|_* \\ &\leq \|f\|_{\mathcal{H}^1} \cdot \|g\|_*, \end{aligned}$$

where we are using the corollary of the $T(1)$ theorem to the effect that $S : BMO \to BMO$. But this last is majorized by $\|f\|_{\mathcal{H}^1}$. Taking the supremum over all such g, we see that $f \in H^1_{\mathrm{Re}}$. [We are using here the fact that an L^1 function that pairs with all BMO functions is in fact an H^1_{Re} function. This assertion, while plausible, is tricky to prove. See [**HAN**].]

But then f has an atomic decomposition: $f = \sum_j \alpha_j a_j$. Applying the Szegö projection to both sides of this last identity, we obtain the desired holomorphic atomic decomposition for f. ∎

Thus we see that, as expected, the duality statement *for holomorphic H^1* implies the holomorphic atomic decomposition for $\mathcal{H}^1$. It is also the case that if one *assumes* the holomorphic atomic decomposition then one may prove that

$(\mathcal{H}^1)^* = BMOA$. In the next subsection we give a précis of an independent proof that $(\mathcal{H}^1)^* = BMOA$.

Holomorphic Duality

We shall quickly indicate the main steps in the proof of the holomorphic version of Fefferman's theorem. Although the argument that we describe holds both on strongly pseudoconvex domains and on domains of finite type in dimension 2, we shall concentrate on the former situation for simplicity. Full details may be found in [**KL1**].

Now fix a strongly pseudoconvex Ω with smooth boundary and defining function $\rho(z)$.

STEP 1: *If $f \in BMOA(\Omega)$ then the expression $|\nabla f|^2|\rho(z)dV$ is a Carleson measure.* The proof of this statement follows classical lines, as in [**FST**]. The key facts that are used are these: (i) that we have sharp size estimates on the Bergman kernel (see [**FEF**] or [**BMS**]), (ii) that the Szegö integral maps L^p onto H^p, $1 < p < \infty$ (see [**PHS**] or the $T(1)$ theorem), and (iii) that Ω satisfies Condition R.

STEP 2: Let $f \in L^1(\partial\Omega)$. Define a maximal function, for z in a tubular neighborhood U of $\partial\Omega$, by

$$M(f)(z) =$$

$$sup \left\{ \frac{1}{|\beta(z_0, r)|} \int_{\beta(z_0,r)} |f(\xi)|\, d\sigma(\xi) : \beta(\pi(z), |\rho(z)|) \subset \beta(z_0, r) \subset \partial\Omega \right\}$$

Here $\beta(p, s)$ stands for the usual non-isotropic balls and $\pi : U \to \partial\Omega$ is orthogonal projection. In this maximal function we are taking the supremum over all boundary balls containing $\beta(\pi(z), |\rho(z)|)$. *This maximal function is will give us a quantitative maximum principal: it controls interior values of u by boundary values in a precise way.*

Now we have:

> *Let u be a non-negative plurisubharmonic function in Ω with boundary function in $L^1(\partial\Omega)$. Then*
>
> $$|u(z)| \leq CM(u)(z)$$
>
> *for $z \in U$.*

This result is due to Hörmander [**HOR5**]. Results similar in spirit may be found in [**STE3**], [**BARK**], [**KRA8**]. The product geometry of a boundary point plays a vital role in the proof. We first majorize $u(z)$ by its mean value over a non-isotropic polydisc. Then we relate the values of u on the polydisc to values on the boundary using (i) harmonic majorization and (ii) classical estimates on the Poisson kernel.

STEP 3: *If $f \in H^1(\Omega)$ and $u \in H^1(\Omega)$ then*

$$|\langle f, u\rangle| \leq C\|f\|_* \cdot \|u\|_{\mathcal{H}^1}.$$

Thus $BMOA \in (H^1)^*$. This result parallels the classical calculation in [**FST**]. We combine Steps 1 and 2 using Green's theorem.

Thus we have established half of the desired conclusion.

STEP 4: *It holds that $(\mathcal{H}^1(\Omega))^* \subseteq S(L^\infty(\partial\Omega))$.* As indicated earlier in the section, this is elementary functional analysis.

STEP 5: *The hypotheses of the $T(1)$ theorem hold for the Szegö kernel.* This is the piece of the puzzle that the folklore wisdom never considered. It also proved to be the most elusive. In particular, one needs to know information about the difference $S(z,\zeta) - S(z,w)$ that is not easily inferred from the standard information about S.

To address this matter, we use the aforementioned device from [**NRSW**] of expressing the Szegö kernel as an integral over normals to the boundary of the Bergman kernel (where precise estimates on the interior are already known thanks to [**FEF**], [**BMS**]).

STEP 6: *It holds that $S(L^\infty) \subseteq BMO$.* This is standard from STEP 5 and the usual techniques as in [**FST**].

The proof that $(\mathcal{H}^1)^* = BMOA$ is complete.

2.4. The Bergman Kernel and Metric. The last twenty years have been witness to a golden age in the study of the ideas of S. Bergman, notably the "kernel function" and the Bergman metric. Bergman's ideas date to the 1920's (see [**BER1**] and references therein; an independent history is provided in [**KRA1**]) and predate those of Carathéodory and Kobayashi/Royden. Hence they serve as an inspiration to much that has been said in this book.

We should like to use this section to give a selective survey of some of this recent activity. This will give us an opportunity to treat some topics that have received scant attention elsewhere, and also to showcase some problems that should perhaps receive more attention. As has been the case in other parts of this book, my method of "selection" is strongly influenced by what I know and by what I have been able to contribute to the field.

We refer the reader to [**KRA1**] and [**KRA7**] for background on the Bergman kernel and metric and for other basic ideas from the function theory of several complex variables that are used here.

One of the papers that opened up modern techniques in the subject was [**DIE**]. In this work Diederich treated the boundary behavior of the Bergman kernel (off the diagonal) on a strongly pseudoconvex domain. He used comparison techniques—particularly that a strongly pseudoconvex boundary point can be closely approximated by a spherical point—to obtain control on $|K(z,\zeta)|$, $|\nabla_z K(z,\zeta)|$, and on $|\nabla^2 K(z,\zeta)|$.

Diederich's ideas served as a model for I. Graham [**GRA1**] who used a very precise form of the "strongly pseudoconvex point is nearly spherical" algorithm to obtain *exact* boundary asymptotics of the Carathéodory and Kobayashi/Royden metrics. Graham's results were further refined in GRA2 and in the thesis of Aladro (see [**ALA1**]).

For me the seminal paper that connected up questions of Bergman geometry with modern powerhouse techniques, such as the $\bar{\partial}$-Neumann problem, was that of N. Kerzman [**KER**]. In that paper he drew attention to the fact that the local hypoellipticity of the $\bar{\partial}$ Neumann problem can be used to obtain direct information about the kernel function itself. [It should be noted here that the crucial formula

$$P = I - \bar{\partial}^* N \bar{\partial} \tag{1}$$

had been noted by Kohn, but not exploited in this context—see [**KRA7**] for details on this matter.] Kerzman's results—in effect that the kernel $K_\Omega(z, \zeta)$ for a strongly pseudoconvex domain Ω is smooth on $\bar{\Omega} \times \bar{\Omega} \setminus$ (diagonal)—had eluded Bergman and others for many years.

While Kerzman developed clever *ad hoc* techniques in order to invoke the $\bar{\partial}$-Neumann problem and extract the desired information from (1), it is possible to use modern ideas from the theory of partial differential equations to get many of his results instantly. Namely, one notices that the Bergman kernel $K(z, \zeta)$ is nothing other than $P(\delta_z)$, where δ_z is the unit mass at z. Now the $\bar{\partial}$- Neumann problem is hypoelliptic (see [**FOLK**] or [**KRA7**]): if f is a function then Nf is smooth wherever f is. This last assertion may be extended to distributions just by formalism. In particular, $N(\bar{\partial}\delta_z)$ is smooth away from z. It follows from equation (1) that so is $K(z, \zeta) = P(\delta_z)$ smooth in ζ *up to the boundary*. The smoothness in z follows from Hermitian symmetry. The *joint* smoothness follows from uniformity of estimates (see [**KRA2**]).

While the symmetries of a complete circular domain (see [**BEB**]), and more generally "transverse symmetries"(see [**BAR**]), can serve to study regularity properties of the kernel on a restricted class of domains, the $\bar{\partial}$- Neumann problem is the only potentially universal tool (that we presently know of) for this task. The technique of comparison domains is promising in its geometric directness, but even on finite type domains in $\mathbb{C}^2$ we do not know what the domains of comparison should be. We should mention that Gebelt [**GEB**] *has* successfully used the technique of comparison domains to study a very restricted class of deformations of ellipsoids. In this work he capitalizes on some important calculations of D'Angelo [**DAN7**].

Fefferman's seminal paper [**FEF**], which proves that biholomorphic mappings of strongly pseudoconvex domains continue smoothly to the boundary, pays direct homage to the paper of Kerzman. Although we now have easier ways to derive Fefferman's theorem, the *techniques* of Fefferman's paper have proved to be of long term significance. If z is a point near the boundary of a strongly pseudoconvex domain Ω then, in suitable local coordinates, Fefferman showed that the Bergman kernel near z is given by the formula

$$K_\Omega(z,\zeta) = \phi(z,\zeta)\cdot K_B(z,\zeta) + \psi(z,\zeta)\cdot \log L(z,\zeta) + \mu(z,\zeta). \qquad (2)$$

Here K_B is the Bergman kernel for the unit ball and $L(z,\zeta)$ is the Levi polynomial (see [**KRA1**]) for background on these ideas). The functions ϕ, ψ, μ are smooth functions that do not vanish near z. Other methods for obtaining expansions for the Bergman and Szegö kernels may be found, for instance, in [**BMS**], [**EPM**], [**HAG**], [**FAK**], and references therein.

It was Robert Greene who first realized the potential of formula (2) for studying Bergman geometry, and it was Klembeck [**KLE**] who made the first steps in the program. Recall that the Bergman metric is given by the complex Hessian of the potential function $\log K_\Omega(z,z)$. Direct (but tedious) calculation using formula (2) shows that the specific nature of ϕ, ψ, μ is of no importance. The upshot is that the holomorphic sectional curvature of the Bergman metric is asymptotically equal to that of the ball as the boundary is approached. It has already been noted in Section 1.9 that this observation gives a direct proof of Bun Wong's theorem: that the only strongly pseudoconvex domain with non-compact automorphism group is (biholomorphic to) the ball.

Greene and Krantz, notably in the papers [**GK2**], [**GK3**], [**GK5**], developed these ideas further. In a nutshell, they established the hard analytic fact that if the boundary of a strongly pseudoconvex domain is varied smoothly then the Bergman kernel also varies smoothly, in the sense that all of the components of formula (2) vary smoothly. More generally, they showed that if the *complex structure* on the closure of a strongly pseudoconvex domain is varied smoothly then the Bergman kernel varies smoothly.

Let us describe briefly why this information can be useful. A question of considerable interest to geometers is this: if Ω is a domain in space and if the complex structure on Ω is smoothly deformed a little bit, then is the resulting complex analytic object also a domain in space? The spirit of this question cuts to the heart of a problem that we do not understand at all: which n-dimensional complex manifolds are imbeddable into $\mathbb{C}^n$ as domains? An obvious necessary conditions of a topological nature is that the manifold be non-compact. But for a simply connected non-compact complex analytic manifold the question seems to be intractable.

An elegant result of Hamilton ([**HAM1**], [**HAM2**]) says this:

THEOREM 2.4.1. *Let $\Omega \subseteq \mathbb{C}^n$ be a strongly pseudoconvex domain with smooth boundary. Let $\hat{\Omega}$ be the same set, thought of as a complex manifold equipped with a complex structure that is a small smooth perturbation of the standard one inherited from $\mathbb{C}^n$. Then $\hat{\Omega}$ can be re-imbedded into $\mathbb{C}^n$ as a subdomain.*

Hamilton's methods are based on the technique of the Nash-Moser implicit function theorem. Greene/Krantz, inspired by a remark of Hörmander, used the stability of the Bergman kernel under perturbation of the complex structure to produce a new proof of Hamilton's theorem. Here is the idea: Let $i : \Omega \to \mathbb{C}^n$ be the identity ($i(z) \equiv z$). Obviously i is holomorphic in the natural complex structure inherited from $\mathbb{C}^n$. But it is not in general holomorphic when thought of as a mapping from $\hat{\Omega}$ into $\mathbb{C}^n$. However, since the perturbed complex structure is very close to the natural structure, it follows that $g \equiv \bar{\partial}_{\hat{\Omega}} i$ is small in the C^1 topology. Here $\bar{\partial}_{\hat{\Omega}}$ is the $\bar{\partial}$ operator associated to the perturbed complex structure. But then the Henkin solution u (for instance) to the equation $\bar{\partial}_{\hat{\Omega}} u = -g$ will be small in the C^1 topology. Finally, the function $i + u$ will be (i) holomorphic in the perturbed complex structure and (ii) still an imbedding because of the C^1 smallness of u. Thus we have produced a holomorphic imbedding of the domain in the perturbed complex structure back into space.

Another elegant application of the Bergman kernel stability is to understanding a different old chestnut: Is it true that any compact, complex n- dimensional Kähler manifold of strictly negative curvature (curvature lying between two negative constants) is covered by the ball? Mostow/Siu [**MSI**] answered this question in the negative by pasting together two complex ellipsoids in a clever way. Related matters were considered from a different point of view in [**GK3**]; in addition, some sufficient conditions for such a manifold to indeed be covered by the ball were provided there.

The idea is this: the holomorphic sectional curvature of the unit ball $B \subseteq \mathbb{C}^n$ is constant. By the stability result, the holomorphic sectional curvature of any domain sufficiently near to B will be negative and bounded from zero. However it follows from results of [**BSW**] that there are domains arbitrarily near the ball that are not biholomorphic to it (indeed, it follows from ideas in [**GK2**] that such domains are generic). By topology, if such a domain were covered by the ball holomorphically then in fact it would be biholomorphic to the ball. That is a contradiction.

One of the deepest ideas to come out of the stability of the Bergman kernel work was a collection of semi-continuity results. Some of these have already been described in Section 1.10. The spirit is that if Ω_0 is a fixed domain and if Ω is a *sufficiently small perturbation* of Ω_0—the degree of smallness depending crucially on Ω_0—then $\mathrm{Aut}(\Omega)$ is a subgroup of $\mathrm{Aut}(\Omega_0)$ in a natural way. Virtually nothing is known about results of this nature for weakly pseudoconvex domains with smooth boundary. Gebelt [**GEB**] has made some progress on the stability of the Bergman kernel for a very restricted class of weakly pseudoconvex domains. It has yet to be seen whether any semi-continuity results for automorphism groups will follow.

Let us now shift gears for a moment and mention a slightly different sort of stability results. Ramadanov's theorem is as follows:

THEOREM 2.4.2. *Let $\Omega_1 \subseteq \Omega_2 \subseteq \cdots$ be domains. Suppose that $\Omega = \cup_j \Omega_j$ is a bounded domain. Then $K_{\Omega_j}(z,\zeta) \to K_\Omega(z,\zeta)$, uniformly on compact subsets of $\Omega \times \Omega$.*

In the proof, the monotonicity is important since one has no additional information about the Ω_j's (such as pseudoconvexity). In [**GK3**], a result *up to the boundary* was proved, in the strongly pseudoconvex case, using an analysis of the $\bar{\partial}$- Neumann problem. In that work, the Ω_j had to converge to Ω in some C^k topology, but the convergence did not have to be monotone.

In fact it is not difficult to see that, in case the Ω_j are pseudoconvex and converge to Ω in such a way that L^2 holomorphic functions on Ω can be approximated by L^2 holomorphic functions on the Ω_j then the conclusion of Ramadanov's theorem is still true. This condition is met if Ω is Runge in each Ω_j. More generally, if Ω has a Stein neighborhood basis then the desired conclusion still holds.

It is natural to conjecture that a version of Ramadanov's theorem should hold if only $\Omega_j \to \Omega$ in the Hausdorff topology. No other hypotheses on the domains, besides boundedness, should be necessary. However Ramadanov [**RAMA2**] has shown that a necessary condition is that there be an L^2 approximation property for Bergman functions as in the last paragraph.

We mention in passing that one may also formulate questions about convergence of the three metrics—Bergman, Carathéodory, and Kobayashi/Royden—on a domain Ω that is the limit of domains Ω_j. Such considerations have proved important, for instance, in the study of semi-continuity of automorphism groups (see, for example [**GK4, GK5**]). For the Kobayashi/Royden metric, matters are straightforward from normal families considerations as long as the domains in question are taut. For the Carathéodory metric (at least for results up to the boundary) the problem quickly reduces to L^∞ approximation theorems and thence to uniform estimates for the $\bar{\partial}$ problem. The strongly pseudoconvex case was handled in [**GK4**]. Some semi-continuity results appear in [**REI**]. Jiye Yu [**YU4**] has made some progress recently on these questions in the case when Ω is unbounded. These results are of particular interest for applications in scaling arguments.

Much of the current interest in the Bergman kernel stems from work of Bell [**BEL1**]-[**BEL3**], Bell/Ligocka [**BEEL**], and Webster [**WEB**] on the study of biholomorphic mappings. Key in these considerations are two ideas: Bergman representative coordinates and Condition R. Representative coordinates have been mentioned elsewhere in this book, but perhaps deserve a bit more detailed treatment at this point.

Fix a domain $\Omega \subset\subset \mathbb{C}^n$ and fix a point $w \in \Omega$. Define n functions

$$\mu_j : z \longmapsto \frac{\partial}{\partial \bar{\zeta}_j} \log \frac{K(z,\zeta)}{K(\zeta,\zeta)}\bigg|_{\zeta=w} \quad , \quad j = 1, \ldots, n.$$

First notice that $K(\zeta,\zeta)$ is positive. That being the case, $K(z,\zeta)$ will be nonvanishing when z and ζ are both close enough to w. So at least the μ_j are well-defined. Obviously, for each j, μ_j is a holomorphic function of z.

To see that the μ_j form a coordinate system it is convenient to use an old trick from [**KOB2**]. Let ψ_1 be the element of the Bergman space of norm 1 so that $\psi(w)$ is real and maximal. Let ψ_2 be the element of the Bergman space such that $\psi_2(w) = 0$, $(\partial/\partial z_1)\psi_2(w)$ is real and maximal. Continue with each of the first partials, and then go to the second and so forth. Then it is easy to verify that the functions $\{\psi_j\}$ form an orthonormal system.

Indeed, if ψ_2 is not orthogonal to ψ_1 then let μ be the orthogonal projection of ψ_2 onto ψ_1. Then $\psi_1 - \mu$ has smaller norm than ψ_1 but the same value at w. Hence ψ_1 was chosen incorrectly. Contradiction.

The $\{\psi_j\}$ form a basis by for the Bergman space by power series considerations about the point w. Now if one expresses the Bergman kernel as

$$K(z,\zeta) = \sum_{\ell=1}^{\infty} \psi_\ell(z)\overline{\psi_\ell(\zeta)}$$

and calculates the Bergman coordinate functions μ_j then it is immediate that the mapping $z \mapsto (\mu_1(z), \ldots, \mu_n(z))$ has invertible Jacobian at w. It follows that $(\mu_1, \ldots, \mu_n)$ forms a coordinate system at w.

The critical fact about the Bergman representative coordinate system is that, when a biholomorphic mapping is expressed in representative coordinates then it becomes linear. We used this fact decisively in Section 1.11. Let us provide the details: The key fact about the Bergman kernel, which is the source of its invariance properties, is the following transformation formula:

PROPOSITION 2.4.3. *Let Ω_1, Ω_2 be domains in $\mathbb{C}^n$. Let $\Phi : \Omega_1 \to \Omega_2$ be a biholomorphic mapping. Then the Bergman kernels K_{Ω_1} and K_{Ω_2} are related by the formula*

$$K_{\Omega_1}(z,\zeta) = det\ Jac_{\mathbb{C}}\Phi(z) K_{\Omega_2}(\Phi(z), \Phi(\zeta)) \overline{det\ Jac_{\mathbb{C}}\Phi(\zeta)}. \tag{3}$$

The proof is standard and may be found, for instance, in [**KRA1**]. Let us calculate Bergman representative coordinates at a point $w \in \Omega_1$ using both sides of (3) and compare the results.

We have

$$\begin{aligned}
\mu_j &= \frac{\partial}{\partial \bar{\zeta}_j}\left(\log \frac{K_{\Omega_1}(z,\zeta)}{K_{\Omega_1}(\zeta,\zeta)}\right)\Bigg|_{\zeta=w} \\
&= \frac{\partial}{\partial \bar{\zeta}_j}\left(\log \frac{\mathcal{J}(z)K_{\Omega_2}(\Phi(z),\Phi(\zeta))\overline{\mathcal{J}(\zeta)}}{\mathcal{J}(\zeta)K_{\Omega_2}(\Phi(\zeta),\Phi(\zeta))\overline{\mathcal{J}(\zeta)}}\right)\Bigg|_{\zeta=w}
\end{aligned}$$

Here we have let $\mathcal{J}$ denote the (holomorphic) Jacobian determinant of the mapping Φ. Simplifying both sides, we find that

$$\begin{aligned}
\mu_j &= \frac{\left(\partial K_{\Omega_1}/\partial\bar{\zeta}_j\right)(z,w)}{K_{\Omega_1}(z,w)} - \frac{\left(\partial K_{\Omega_1}/\partial\bar{\zeta}_j\right)(w,w)}{K_{\Omega_1}(w,w)} \\
&= \sum_\ell \frac{\partial \bar{\Phi}_\ell}{\partial \bar{\zeta}_j}(w) \cdot \left[\frac{\left(\partial K_{\Omega_2}/\partial\bar{\zeta}_\ell\right)(\Phi(z),\Phi(w))}{K_{\Omega_2}(\Phi(z),\Phi(w))} - \frac{\left(\partial K_{\Omega_2}/\partial\bar{\zeta}_\ell\right)(\Phi(w),\Phi(w))}{K_{\Omega_2}(\Phi(w),\Phi(w))}\right].
\end{aligned}$$

If we let M^1, M^2 denote the vectors of the n representative coordinates on Ω_1 and Ω_2 respectively, then our equation may be written as

$$M^1(w) = \overline{\mathrm{Jac}_{\mathbb{C}}\Phi}(w) M^2(\Phi(w)) \, .$$

By Hermitian transposition, we find that the (unconjugated) Jacobian matrix $\mathrm{Jac}_{\mathbb{C}}\Phi(w)$ represents Φ by linear operation from M^1 to M^2.

We repeat the crucial fact that the neighborhood of w on which the Bergman representative coordinates are valid is, in general, indeterminate. The process of differentiation eliminates any possible ambiguity arising from the definition of the logarithm; *the only obstruction to global definition of Bergman representative coordinates is the vanishing of the Bergman kernel.*

Considerations such as these made the Lu Qi-Keng conjecture of particular interest. In raw form, the conjecture is that the Bergman kernel of a bounded domain never vanishes. Explicit calculation shows that the Bergman kernel $(1/\pi)(1/(1-z\bar{\zeta})^2)$ never vanishes. It follows from the Riemann mapping theorem, and the transformation formula (3), that the Bergman kernel of a bounded, simply connected domain in $\mathbb{C}$ will never vanish. Although there is no Riemann mapping theorem in several complex variables, one can still explicitly calculate the Bergman kernel for the ball and polydisc and see that these are non-vanishing.

The Bergman kernel for the annulus was studied by M. Skwarczynski [**SKW**] and was seen to vanish at some points. It is shown in [**SUY**] that that if $\Omega \subseteq \mathbb{C}$ is a multiply connected domain with smooth boundary then K_Ω must vanish—this is proved by an analysis of differentials on the Riemann surface consisting of the double of Ω. By using the easy fact that the Bergman kernel for a product domain is the product of the Bergman kernels (exercise), we may conclude that any domain in $\mathbb{C}^2$ of the form $A \times \Omega$, where A is multiply connected, has a Bergman kernel with zeros. The Lu Qi-Keng conjecture can thus be reformulated as

> **CONJECTURE:** A topologically trivial domain in $\mathbb{C}^n$ has non-vanishing Bergman kernel.

It is known ([**GK2**], [**GK3**]) that a domain that is C^∞ sufficiently close to the ball in $\mathbb{C}^n$ has non-vanishing Bergman kernel. Also, if a domain Ω has Bergman kernel that is bounded from zero (and satisfies a modest geometric condition) then all nearby domains have Bergman kernel that is bounded from zero. Thus it came as a bit of a surprise when in [**BOA**] it was shown that there exist topologically trivial domains—even ones with real analytic boundary and satisfying all reasonable additional geometric conditions—for which the Bergman kernel has zeros. It is safe to say that at this point we do not know what the Lu Qi-Keng conjecture should say, or if it will be true for any reasonable class of domains.

The paper [**WIE**] contributed interesting ideas to Boas's negative resolution of the Lu Qi-Keng conjecture. In fact he constructs a domain in $\mathbb{C}^2$ for which the Bergman space has positive, finite dimension. He shows that in one complex variable this can never happen—the dimension of the Bergman space of a planar domain is always either zero or infinite.

Next we turn to some recent developments about the size and shape of the Bergman kernel. We shall only touch on some high spots, and give many references for further reading. In the paper [**NRSW**], the authors develop a scaling technique to derive size estimates for the Szegö kernel and its derivatives on a finite type domain in $\mathbb{C}^2$. The idea (which harkens back to some of our major themes in Section 1.3) is that the natural geometry near a boundary point P of type m is the geometry of a bidisc of poly-radius $(\delta, \delta^{1/m})$. Using non-isotropic scaling, one can blow this polydisc up to the standard polydisc $D^2(0,1)$. This blowup map also blows up a neighborhood of P to a "model domain." Thus, up to an error, the Bergman kernel of the original domain Ω near P can be thought of as the Bergman kernel of the model pushed back to Ω by way of the scaling map.

There is an important technical catch in this process. A certain patching argument must be made to mediate between the local geometric construction just described and the fact that the Szegö kernel is a global construct. This patching argument is effected with the $\bar{\partial}$- Neumann problem. If one scales the $\bar{\partial}$ problem in the most naive way, then in fact an unacceptable error of size $\log(1/\delta)$ arises. To circumvent this difficulty, an important localization result of Catlin [**CAT1**] must be exploited. In effect it says that the regularity of the $\bar{\partial}$-Neumann problem—in particular the constants in the estimates—near a point of finite type depends only on local data. We can say no more about it here.

The next step in the program of [**NRSW**] is to relate the Szegö kernel to the Bergman kernel. Perhaps a word about the epistemology of this connection is in order. On the ball in $\mathbb{C}^n$, the Bergman and Szegö are very similar in form:

$$S(z,\zeta) = c_n \frac{1}{(1 - z \cdot \bar{\zeta})^n} \qquad\qquad K(z,\zeta) = C_n \frac{1}{(1 - z \cdot \bar{\zeta})^{n+1}}.$$

It appears that the Bergman kernel is the derivative of the Szegö kernel. This is plausible for the following reason: on a reasonable domain, the functions that are C^1 on $\bar{\Omega}$ and holomorphic on the interior are dense in both the Bergman space and the space H^2 (see [**HAS**] and [**BEA**])—this holds, for instance, on smooth strongly pseudoconvex domains and on finite type domains in $\mathbb{C}^2$. [For H^2 approximation there is a more general result in [**BEA**]; the Bergman space approximation question is closely related to Bell's Condition R from mapping theory as noted in [**BAR**].] Thus one might expect to apply Stokes's theorem to the Szegö integral and obtain something approximating the Bergman integral (or vice versa). This is, in effect, the program that is carried out in [**NRSW**].

Without introducing the minefield of (necessary) notation that appears in [**NRSW**], let us just say a few words about the nature of the estimates that are obtained. For specificity, we shall concentrate our remarks on the Szegö kernel because this kernel is a bit simpler to treat. Now Ω is a domain of finite type in $\mathbb{C}^2$. If z, ζ are two points of $\partial\Omega$ then we may imagine a smallest nonisotropic ball $b_{z,\zeta}$ that contains both z and ζ. Of course we refer here to the balls that arise naturally from the complex structure (Sections 1.5, 1.6 gave an intrinsic way to think about these balls). The normalization in [**NRSW**] is that a non-isotropic ball of radius δ has size about δ in complex tangential directions and size about δ^m in the complex normal direction (where m is the type of the point).

Then we have

$$|S(z,\zeta)| \lesssim \frac{1}{\sigma(b_{z,\zeta})}.$$

Here σ is $2n-1=3$ dimensional area measure on $\partial\Omega$.

To understand this estimate, it is instructive to look at a specific example that we already understand. Let $\Omega = B$ be the unit ball in $\mathbb{C}^2$. Then the Szegö kernel for this domain is

$$S(z,\zeta) = c \cdot \frac{1}{(1 - z \cdot \bar{\zeta})^2}.$$

Now we know that the expression $|1 - z\cdot\bar{\zeta}|$ measures the *complex normal component* of the distance from z to ζ. If the smallest non-isotropic ball that contains z, ζ has radius r, then the volume of the ball is $r^2 \cdot r \cdot r = r^4$. But then the normal piece of the distance from z to ζ is $\sim r^2$ and we see that the denominator of the formula for S is of size r^4. So, at least for the unit ball, we can understand the estimate from [**NRSW**].

In independent work, McNeal [**MCN1**] has carried out a program that is quite similar to the one in [**NRSW**], and is in some ways more elegant. We mention in particular that McNeal [**MCN1**], [**MCN3**] has introduced a parameter τ which is rather natural for measuring the geometry that we have been discussing.

Part of the work in [**NRSW**] and [**MCN1**] is to also understand derivatives of the kernel (the point of view here is to develop the machinery for later applications of the $T(1)$ theorem). If z, ζ are points in $\partial\Omega$ then, following the notation of [**NRSW**], we let $\rho(z,w)$ denote the non-isotropic distance normalized so that a path of Euclidean length δ in a complex tangential direction is declared to have length δ and a path of Euclidean length δ in a complex normal direction is declared to have length δ^m, where m is the type (this normalization is consistent with the aforementioned normalization of balls—indeed the balls may be defined using this notion of distance). If X_z^j denotes a tangential derivative of order j and X_ζ^k denotes a tangential derivative of order k then [**NRSW**] contains the estimate

$$|X_z^j X_\zeta^k S(z,\zeta)| \le C_{j,k} \frac{\rho(z,\zeta)^{-j-k}}{\sigma(b_{z,\zeta})}. \tag{4}$$

The papers [**NRSW**] and [**MCN1**] also contain parallel estimates for the Bergman kernel; these are a bit more complicated to formulate because of the intervention of the boundary.

Refer to Section 2.3 for a précis of the $T(1)$ theorem and singular integral theory. A more thorough treatment is in [**CHR1**]. In order to see that an operator is bounded on L^p, one must check that the kernel and its derivatives decay at a rate that is suggested by scaling. Such estimates are, of course, the content of line (4). The weak boundedness condition also follows from scaling. Finally the condition that $T(1) \in BMO$ comes for free in the present circumstance because the Szegö projection of the constant function 1 is the constant function 1 itself. Thus the $T(1)$ theorem tells us that the Szegö projection on a finite type domain in $\mathbb{C}^2$ maps L^p to itself, $1 < p < \infty$. More recent developments in the $T(1)$ theory tell us that the Szegö projection also maps BMO to BMO and H^1 to H^1. We shall say no more about these matters here, but refer the reader to [**FTW**] and [**HJMT**].

Some of the most exciting recent developments in Bergman theory concern convex domains in any dimensions. In a nutshell, the geometry for these domains is much simpler than for a generic domain. We have already mentioned in Sections 1.2 and 1.3 the fact that line type and variety type are equal on a convex domain (see [**MCN6**], [**BOS**], and [**YU2**]). Moreover, the more sophisticated notion of multi-type may also be measured with lines when the domain is convex. Part and parcel of the circle of ideas that led to these insights are some new estimates for the Bergman kernel. In the paper [**MCN2**] McNeal introduces a class of "decoupled domains", in any dimension, for which useful estimates of the Bergman kernel and metric can be obtained; the use of the word "decoupled" denotes the independent manner in which the variables are hypothesized to appear in the defining function.

Let us just say roughly what the estimates are. To any convex boundary point of finite type (see [**MCN6**]), McNeal associates a non-isotropic analytic object A that measures complex extent in an $(n-1)$ dimensional coordinate frame for the tangential directions. This complex extent may be measured using just complex lines. Then McNeal can obtain sharp upper bounds for the Bergman kernel $K(z,\zeta)$ in terms of powers of $A(z,\zeta)$. He can also obtain sharp lower bounds, but only on the diagonal. It is probable that sharp lower bounds on all of $\Omega \times \Omega$ lie rather deep since they would address (among other things) the Lu Qi-Keng conjecture.

Let us close this section with a few remarks about localization of analytic questions. We are beginning to develop a good working understanding of local analysis near finite type boundary points—especially in dimension two. When our analysis is successful, as in the program of [**NRSW**], the passage to global data is effected by way of (for instance) $\bar{\partial}$ techniques. Along with sheaf theory and (what is the same thing) Cousin problems, this is our chief means of passing from local to global information. For some types of problems these techniques are inadequate. For instance, when does estimation of the Kobayashi metric depend only on local boundary data? Is it not easy to picture a situation in which the calculation of the Kobayashi metric near a boundary point depends on discs

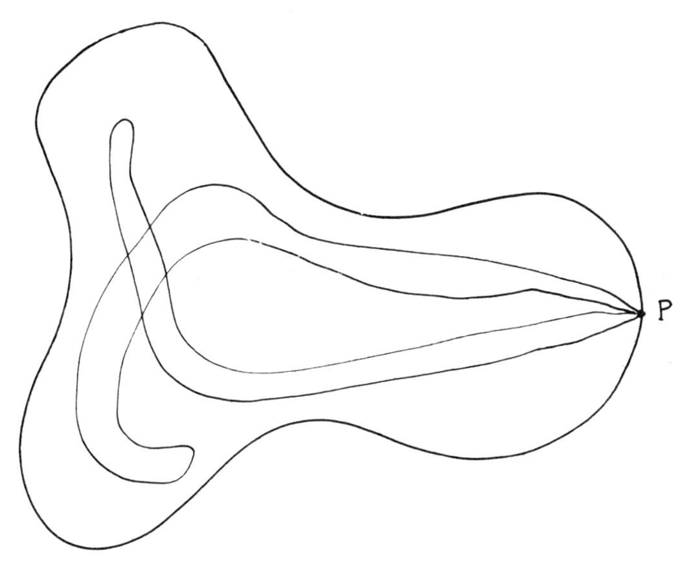

FIGURE 2.3

that live in the entire domain (Figure 2.3)? The most pervasive technique for localizing the Kobayashi metric near a boundary point P is to utilize peaking functions at P (when they exist). But we do not know precise conditions for the existence of such functions; what we do know seems to be independent of $\bar{\partial}$ ideas or sheaf theory ideas.

We record here two recent results about the localization of Bergman kernel and metric quantities. Both of these should prove useful, at least philosophically, in future analysis.

PROPOSITION 2.4.4 (MCNEAL [**MCN4**]). *Let Ω be a smoothly bounded, pseudoconvex domain in $\mathbb{C}^n$ and $P \in \partial\Omega$. Suppose that $D \subseteq \Omega$ is a smoothly bounded domain such that $\partial D \cap \partial\Omega$ contains a relative neighborhood of p in $\partial\Omega$. Finally, let U be a neighborhood of p in $\mathbb{C}^n$ such that*

$$\partial(U \cap \Omega) \subseteq \bar{D} \qquad \text{and} \qquad \partial U \setminus {}^c\Omega \subseteq D.$$

Then, for any tangent vector X and $z \in U \cap \Omega$, the Bergman metric B_Ω satisfies

$$B_\Omega(z, X) \lesssim B_D(z, X) \lesssim B_\Omega(z, X).$$

Also the kernel K_Ω satisfies

$$K_\Omega(z, z) \lesssim K_D(z, z) \lesssim K_\Omega(z, z)$$

for all $z \in U \cap \Omega$ and all vectors $X \in \mathbb{C}^n$.

What is more, if D is a differential operator containing the same number of holomorphic as anti-holomorphic derivatives, then one can localize the estimates on DK in a similar fashion.

On the one hand, McNeal's result is just the sort of localization theorem that one would wish for. McNeal's techniques cover not only the curvature but any derivative of the kernel in which the number of holomorphic and antiholomorphic derivatives are in balance. But the results are only valid on the diagonal.

For comparison, here is a result from [**GK7**] that is valid on $\Omega \times \Omega$. However it involves some imprecision and some extra hypotheses.

PROPOSITION 2.4.5 (GREENE/KRANTZ [**GK7**]). *Let $\Omega \subseteq \mathbb{C}^2$ be a smoothly bounded pseudoconvex domain. Fix $P \in \partial\Omega$ a point of finite type m. Fix an integer $k > 0$. Let $A \subseteq \Omega$ be a subdomain with the following properties:*

1: $\overline{A} \cap \partial\Omega = \{P\}$;

2: *A has smooth boundary near P;*

3: *∂A osculates $\partial\Omega$ to some finite order ℓ at P;*

Then there is a positive integer $N = N(k, \ell, m)$ such that if Ω' is another domain that osculates $\partial\Omega$ to order N at P and if $A \subseteq \Omega'$ then

$$\sup_{z,w \in A} \left| \nabla_z^\alpha \nabla_{\overline{w}}^\beta \left(K_\Omega(z,w) - K_{\Omega'}(z,w) \right) \right| \leq M$$

for some finite positive constant M, for all multi-indices α and β satisfying $|\alpha| + |\beta| \leq k$, and for all $z, w \in A$,

The idea of the proof is simple, and is based on Fefferman's approach in [**FEF**]. Fefferman obtains an asymptotic expansion for the Bergman kernel of a strongly pseudoconvex domain Ω by approximating a strongly pseudoconvex boundary point to fourth order by the biholomorphic image U of a ball. The fourth order approximation is crucial, for it results in an error term that is supported on the set theoretic difference $\Omega \setminus U$. The smallness of the wedge (see Figure 2.4) is decisive in controlling the error. *Note that one can predict in advance how asymptotically small this wedge needs to be by simply examining the singularity of the Bergman kernel for the ball.*

Now [**NRSW**] teaches us the size of the singularity of the Bergman kernel near the finite type boundary point P. We select the number N to ensure that the wedge that supports the error is asymptotically small enough to kill the singularity of the error in Fefferman's construction in this circumstance. *One caveat:* In order to avoid the difficult integrations by parts in Fefferman's paper, we restrict the variables to lie in the asymptotic region A. See Figure 2.5. We refer the reader to [**GK7**] for details.

2.5. Designing Hardy Spaces.
Introductory Remarks

In this section we give a detailed example of the philosophy of designing function spaces—inspired, of course, by the classical ones—designed for specialized applications. We begin with a review of the classical setup.

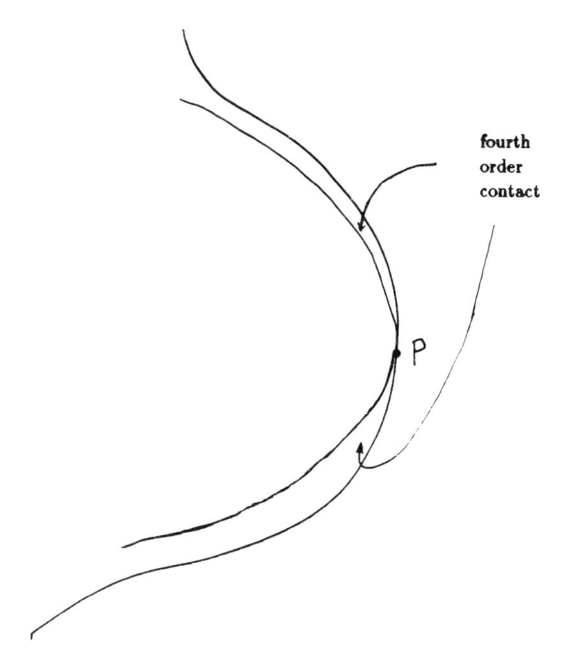

FIGURE 2.4

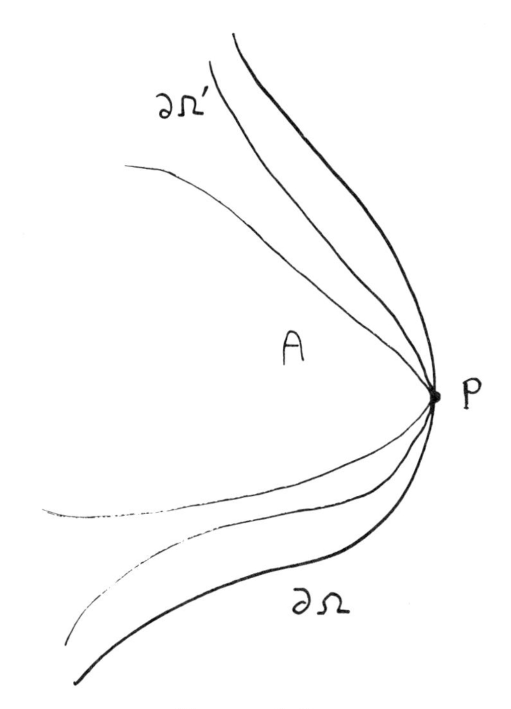

FIGURE 2.5

Recall the classical Hardy space $H^1(D)$ on the disc $D \subseteq \mathbb{C}$:

$$H^1(D) = \{f \text{ holomorphic on } D : \sup_{0<r<1} \int_0^{2\pi} |f(re^{i\theta}| \, d\theta < \infty\}.$$

Then each element of H^1 has radial boundary limits for almost every $\theta \in [0, 2\pi)$ (see Sections 1.5, 1.6 on the boundary behavior of holomorphic functions). Call the boundary function f^*. It is clear, from Fatou's lemma for instance, that $f^* \in L^1([0, 2\pi))$. We may identify H^1 with a proper subset of L^1 by the mapping

$$H^1 \ni f \mapsto f^* \in L^1. \tag{1}$$

One sees from the elementary theory of Fourier series that, because

$$f(z) = \sum_{j=1}^{\infty} a_j z^j,$$

therefore the function f^* can have non-zero Fourier coefficients only corresponding to non-negative frequencies. In particular, the mapping (1) cannot be onto. For instance the functions $e^{ij\theta}, j < 0$ cannot arise as functions f^* for any $f \in H^1$.

Let $f \in L^1([, 2\pi))$ have positive real part. Then we denote the Poisson integral of f to the disc by $u(re^{i\theta})$. We select the harmonic conjugate v that vanishes at the origin. Then $h = u+iv$ is holomorphic on D. What is more, u is non-negative so that $H = e^{-u-iv}$ is a bounded holomorphic function on D. It follows that v has a boundary function (since H does) that we call $v^* \equiv \tilde{f}$. The function $\tilde{f}$ is called *the Hilbert transform* of f. Notice that the construction of this paragraph extends to all real valued $f \in L^1$ by real linearity, and to all complex valued $f \in L^1$ by complex linearity.

In case $\tilde{f}$ is in L^1 then it is clear that $h \in H^1$. Since h is uniquely determined by f, we say in this circumstance that the function f is in *real* H^1. We write $f \in H^1_{\text{Re}}$. Thus

$$H^1 \ni f \mapsto \text{Re}\, f$$

is a one-to-one identification of H^1 with H^1_{Re}.

In the preceding calculations, one may find an integral kernel that produces v from f—see [**KAT**]. It is essentially the imaginary part of the Cauchy kernel on the disc. Indeed, we have that

$$v(re^{i\theta}) = \frac{1}{2\pi} \int_{-\pi}^{\pi} f\big(e^{i(\theta-\psi)}\big) \frac{2r \sin \psi}{1 - 2r \cos \psi + r^2} \, d\psi.$$

As $r \to 1^-$ we are led to consider the *singular integral kernel* $K(\psi) - \cot(\psi/2)$. Indeed, it can be shown that

$$Hf \equiv f^*(e^{i\theta}) = P.V. \int_{-\pi}^{\pi} f(e^{i\psi}) K(\theta - \psi) \, d\psi.$$

Here $P.V.$ denotes the *Cauchy principal value*, and means that we interpret the integral as $\lim_{\epsilon \to 0^+} \int_{|\psi|>\epsilon}$. The operator H is called the *Hilbert transform*. Since the kernel K is not in L^1, the integral can only be interpreted in this special sense. The operator H is a *singular integral operator.*

It has been known for some time (see [**KAT**] and [**STE1**]) that the operator H is bounded on $L^p, 1 < p < \infty$, but neither on L^1 nor on L^∞. In view of the preceding discussion, it is reasonable to consider the space of $f \in L^1$ such that $Hf \in L^1$. This turns out, of course, to be H^1_{Re}.

Thanks to Stein and Weiss [**STW**], there is a theory of real variable Hardy spaces on any $\mathbb{R}^N$. The role of the Hilbert transform is played by the set of *Riesz transforms*: set $K_j(x) = c_N x_j/|x|^{N+1}$. Then

$$R_j(f) \equiv P.V. \int_{\mathbb{R}^N} K_j(x-t)f(t)\,dt.$$

We use the Cauchy principal value again because the kernels K_j do not lie in L^1 and the integrals can only be understood in this special sense. The Riesz transforms are canonical in the sense that they form the only N-tuple of operators that commute with translation, dilation, and rotation. We say that a function $f \in L^1(\mathbb{R}^N)$ lies in H^1_{Re} if $R_j f \in L^1$ for $j = 1, \ldots, N$.

Recall that Calderón-Zygmund theory teaches us that a singular integral is bounded on $L^p, 1 < p < \infty$. It follows then from the definition that $H^p_{\text{Re}} = L^p, 1 < p < \infty$.

The space H^1_{Re} is natural in many senses. For example, the elements of this space can be thought of, in a reasonable sense, as boundary functions for a holomorphic function of several complex variables on a wedge (see [**CAR**]). Calderón-Zygmund singular integral operators are bounded on elements of H^1_{Re}. In a natural sense, H^1_{Re} is the largest subspace of L^1 on which singular integrals behave boundedly.

However the definition in terms of Riesz transforms is somewhat limiting. It is quite difficult to study directly the behavior of an integral operator on H^1 using the Riesz transform definition. And many of the more interesting analytic problems take place on the boundary of a domain or, more generally, on a manifold. Thus one wishes to free the theory from the translation structure of $\mathbb{R}^N$ and from the dependence on a particular N-tuple of kernels. [One should add here that Uchiyama [**UCH**] has determined precisely *which* N-tuples of singular integral operators characterize H^1_{Re}. This is a matter of considerable interest, but we wish to transcend even that result.]

As an intermediate step, we mention the Burkholder-Gundy-Silverstein theorem. First proved on the disc by probabilistic methods, the statement here is on $\mathbb{R}^N$ (see [**FST**] for proof and background). Recall that the Poisson kernel for the upper half space $\mathbb{R}^{N+1}_+$ is given by

$$P_y(x) = c_N \frac{y}{(x^2+y^2)^{N/2}}.$$

If f is a function on $\mathbb{R}^N = \partial\mathbb{R}^{N+1}_+$ with reasonable decay at infinity then we may define

$$P_y f(x) = P_y * f.$$

Set

$$f^+(x) = \sup_{y>0} |P_y f(x)|.$$

Now the theorem is

THEOREM 2.5.1. *Let $f \in L^1(\mathbb{R}^N)$. If $f^+ \in L^1(\mathbb{R}^N)$ then $f \in H^1_{\text{Re}}$. The converse is true as well.*

This was the first "real variable" characterization of the Hardy space H^1. It makes no reference to holomorphic function theory. It *does* depend on the theory of harmonic functions. But it set the tone for what is now to follow.

We next present two intrinsic characterizations of H^1_{Re} that lend themselves naturally to fairly general contexts. The proof that these characterizations are equivalent to the original definition is deep, and we cannot treat it here. However see [**FST**]. For the first characterization, fix a function $\phi \in C_c^\infty(\mathbb{R}^N)$, normalized so that $\int \phi \, dx = 1$. Set $\phi_\epsilon(x) = \epsilon^{-N}\phi(x/N)$. If $f \in L^1_{\text{loc}}(\mathbb{R}^N)$ then we define

$$f^*(x) = \sup_{\epsilon>0} |f * \phi_\epsilon(x)|.$$

THEOREM 2.5.2 (FEFFERMAN/STEIN). *Let $f \in L^1(\mathbb{R}^N)$. Then $f \in H^1_{\text{Re}}$ if and only if $f^* \in L^1(\mathbb{R})$.*

It turns out that the most flexible characterization, which is derived from a variant of the maximal function definition just given, is as follows. We first define a measurable function $a(x)$ to be a *1-atom* if

1: $\operatorname{supp} a \subseteq B(x,r)$ for some $r > 0$;

2: $|a(x)| \leq \dfrac{1}{|B(x,r)|}$;

3: $\displaystyle\int a(x)\, dx = 0.$

Of course we have seen atoms at several junctures earlier in the book.

THEOREM 2.5.3. *Let $f \in L^1(\mathbb{R}^N)$. Then $f \in H^1_{\text{Re}}$ if and only if there are a sequence $\{\lambda_j\} \in \ell^1$ and atoms a_j such that*

$$f = \sum_j \lambda_j a_j.$$

The "atomic characterization" of H^1 means, in effect, that one can check boundedness of integral operators on H^1 by just checking on an atom. This is surprisingly straightforward. It is unfortunately *not* the case that singular integrals map atoms to atoms. However there are slightly more general objects, called *molecules*, which are preserved under singular integral operators. We shall not discuss molecules here, but refer the reader to [**TAW**].

We mention in passing that there is a real variable H^p theory, $0 < p < 1$, which lends itself to all three characterizations that we have presented. However the boundary function of even a classical H^p function on the disc is not integrable. In the setting of $\mathbb{R}^N$ an element of H^p_{Re} must be interpreted as a distribution. This fact leads to unpleasant technicalities that are best avoided in an exposition such as this one. Therefore we shall confine our attention primarily to H^1 type spaces.

In addition to behaving well under the action of singular integral operator, H^1_{Re} also behaves nicely under the action of the fractional integral operators given by the kernels $I_\alpha(x) = c_N|x|^{-N+\alpha}, 0 < \alpha < N$. In particular, I_α maps H^1_{Re} to $L^{N/(N-\alpha)}$ and, more generally, H^p_{Re} to H^q_{Re} with $1/q = 1/p - \alpha/N$. What we wish to establish here is that the real variable Hardy spaces serve as a natural substitute for L^p when $p \leq 1$: singular integral operators and fractional integral operators, the two most fundamental operators in all of analysis (the Hilbert transform is a singular integral; the Newtonian potential is a fractional integral) preserve the H^p spaces rather than the L^p spaces at the level $p \leq 1$.

In spite of their naturality, the real variable Hardy spaces are not suited to many of the most standard situations in analysis—especially in partial differential equations. For instance, if ψ is a smooth cutoff function and $f \in H^1_{\mathrm{Re}}$ then it is not generally the case that $\psi \in H^1_{\mathrm{Re}}$. [Check this assertion when f is an atom.] Second, if Φ is a diffeomorphism of $\mathbb{R}^N$ and if $f \in H^1_{\mathrm{Re}}$ then it is not generally the case that $f \circ \Phi$ lies in H^1_{Re}. To see this, notice that composition with a diffeomorphism does not preserve the mean value condition for atoms. Both of these shortcomings are serious problem if we want to define H^1_{Re}, using a partition of unity, on a manifold.

A perhaps more profound shortcoming of the real variable Hardy spaces, as defined, is that they are not preserved under the action of, say, zero order pseudodifferential operators. Indeed we could not expect that they would be, for H^1_{Re} is translation invariant while pseudodifferential operators generically are not.

To address the indicated problem, we have the *local Hardy spaces* of Goldberg [**GOL**]. These may be defined using a local theory of Riesz transforms, but these are beside the point for our purposes. Thus we concentrate on the other two, and especially on the third, characterization.

With ϕ a C^∞_c testing function of unit mass, ϕ_ϵ as before, and $f \in L^1_{\mathrm{loc}}$, we set

$$f^+_{\mathrm{loc}} = \sup_{1>\epsilon>0} |f * \phi_\epsilon(x)|.$$

We say that $f \in h^1_{\mathrm{loc}}$ if $f^+_{\mathrm{loc}} \in L^1$. Of course $H^1 \subseteq h^1_{\mathrm{loc}}$. To see that the containment is strict, check that any C^∞_c function is in H^1_{loc}.

The atomic characterization of h^1_{loc} is as follows. Define a measurable function $a(x)$ to be a *local* 1-*atom* if

1: $\mathrm{supp}\, a \subseteq B(x, r)$ for some $r > 0$;

2: $|a(x)| \leq \dfrac{1}{|B(x,r)|}$;

3: $\int a(x)\,dx = 0$ if $r \le 1$; no moment condition is imposed otherwise.

Elements of local h^1 have atomic decompositions in terms of local atoms. The lack of a moment condition on atoms with large support is the concrete realization of the "locality" of the space. More precisely, if f is any function on $\mathbb{R}^N$ then define let $\mathcal{K}f(x) = |x|^{-N+2} f(x/|x|^2)$. This is the Kelvin reflection of f. Recall (see [**KRA1**]) that harmonic functions are preserved under Kelvin reflection. Goldberg's theorem [**GOL**] is

THEOREM 2.5.4. *Let $f \in L^1(\mathbb{R}^N)$. Then $f \in H^1_{\mathrm{Re}}$ if and only if $f \in h^1_{\mathrm{loc}}$ and $\mathcal{K}f \in h^1_{\mathrm{loc}}$.*

The local Hardy spaces also satisfy the following convenient properties:

THEOREM 2.5.5. *We have that*

1: *h^1_{loc} is closed under multiplication by cutoff functions;*

2: *h^1_{loc} is closed under composition with diffeomorphisms;*

3: *h^1_{loc} is preserved under order zero pseudodifferential operators.*

For our purposes in this section an order zero pseudodifferential operator is a Fourier integral operator

$$f \mapsto \int e^{ix\cdot\xi} \sigma(x,\xi) \hat{f}(\xi)\, d\xi,$$

where $p(x,\xi)$ is an order zero symbol: that is, $\sigma(x,\xi)$ is a jointly C^∞ function that satisfies

$$\left| \left(\frac{\partial}{\partial x}\right)^\alpha \left(\frac{\partial}{\partial \xi}\right)^\beta \sigma(x,\xi) \right| \le C_{\alpha,\beta}(1+|\xi|)^{-|\beta|}.$$

In the theory of elliptic operators, which is the main focus of the present discussion, all of the parametrices that we shall study shall be written as the composition of fractional integral operators, pseudodifferential operators of order zero, and Hilbert integral operators (more on these below—see also [**KRA7**]).

In an ideal world, an analyst would encounter a problem and produce a calculus of integral operators that serves to study *that particular problem*. This program has been carried out in a limited number of cases: The classical calculi of singular integrals, fractional integrals, etc. were (in retrospect) designed to study the Laplace operator and allied objects. Special calculi have been designed for the study of local solvability issues (see [**BEF**]). In [**GRS**] a calculus was developed to study the $\bar\partial$-Neumann problem on a strongly pseudoconvex domain. Nagel and Stein [**NAS2**] developed a calculus for studying certain subelliptic problems. Geller [**GEL**] produced a calculus for treating certain real analytic hypoellipticity questions.

We would like to focus here on a different program: that of producing *function spaces*. The standard function spaces in analysis—those of Lebesgue, Lipschitz, Sobolev, Besov, etc.—were invented for the study of the Laplace operator. Their structure reflects the symmetries of the Laplacian, and of Euclidean space. We present here a construction of function spaces designed for the regularity theory of an elliptic boundary value problem on a bounded domain in space. Note that, when working on a domain, we no longer have any of the usual spatial homogeneities. Thus a new point of view is required.

The Main Problem

Consider the following two questions:

Question 1 Let Ω be an (appropriate) domain in $\mathbb{R}^N$. What are the possible (natural) notions of $H^p(\Omega)$ that generalize the usual Hardy spaces $H^p(\mathbb{R}^N)$?

Now let $\mathbf{G}$ and $\tilde{\mathbf{G}}$ be the Green's operators for the Dirichlet problem

$$\begin{aligned} \Delta u &= f \quad \text{in} \quad \Omega \\ u\big|_{\partial\Omega} &= 0 \end{aligned}$$

and the Neumann problem

$$\begin{aligned} \Delta u &= f \quad \text{in} \quad \Omega \\ \frac{\partial u}{\partial \vec{n}}\Big|_{\partial\Omega} &= 0 \end{aligned}$$

respectively. Our next, and more focused, question is:

Question 2 In the context of the relevant $H^p(\Omega)$, can one obtain the boundedness of $f \to \frac{\partial^2 \mathbf{G}}{\partial x_i \partial x_j}(f)$, and $f \to \frac{\partial^2 \tilde{\mathbf{G}}}{\partial x_i \partial x_j}(f)$?

This sort of boundedness is known when either (i) $1 < p < \infty$ or (ii) the domain is all of space.

Now we fix a bounded, connected domain $\Omega \subset \mathbb{R}^N$. Assume for simplicity that Ω has smooth boundary (many of the results are valid if only the domain has Lipschitz boundary). Let $p > 0$. We define two h^p spaces.

DEFINITION 2.5.6. We set $h^p_r(\Omega)$ equal to the collection of those distributions f on Ω such that there is an $F \in h^p(\mathbb{R}^N)$ whose restriction to Ω equals f.

DEFINITION 2.5.7. We define $h^p_z(\Omega)$ to be those f on Ω that arise by restricting to Ω the distributions F in $h^p(\mathbb{R}^N)$ such that $F \equiv 0$ in ${}^c\overline{\Omega}$.

We make the following elementary observations:

1: $h^p_r(\Omega) = h^p_z(\Omega) = L^p(\Omega) \quad , \quad 1 < p < \infty$;

2: $h^p_z(\Omega) \subset h^p_r(\Omega)$ but $h^p_z(\Omega) \neq h^p_r(\Omega)$ when $0 < p \leq 1$.

Details of the matters discussed here appear in [**CKS1, CKS2**].

Principal Results

Let us assume for the moment that $\Omega \subset \mathbb{R}^N$ is bounded and smooth. We will use the notation from our discussion of the Dirichlet and Neumann problems for the Laplacian.

Now we state the primary theorems. For motivation note that the fundamental solution of the Laplacian $\Gamma(x) = c_n|x|^{-N+2}$ is a fractional integral operator. Estimating the L^p norm of the solution $u = \Gamma * f$ of the Laplace equation is an elementary exercise in (absolutely convergent) integration theory. The critical estimate, which is not elementary, is to study $\partial^2 u/\partial x_j \partial x_k$. This problem gives rise to the singular integral

$$f \mapsto \left(\frac{x_j x_k}{|x|^{N+2}} \right) * f.$$

That such an integral operator is bounded on $H^p(\mathbb{R}^N), 0 < p < \infty$, is part of the classical theory of Fefferman/Stein/Weiss (see [**FST**], [**STW**], [**KRA13**]). We wish to study the analogue of this problem on a smoothly bounded domain.

THEOREM 2.5.8. *Let* $\mathbf{G}$ *be the Green's operator of the Dirichlet problem for the Laplacian on* Ω. *Then the operators*

$$\frac{\partial^2 \mathbf{G}}{\partial x_j \partial x_k}$$

originally defined on $C^\infty(\overline{\Omega})$, *can be extended as bounded operators from* $h^p_r(\Omega)$ *to* $h^p_r(\Omega)$, *for* $1 \le j, k \le N$ *and* $\frac{N}{N+1} < p \le 1$.

This assertion fails when $p \le \frac{N}{N+1}$. However if we assume that the data belongs to the smaller space $h^p_z(\Omega)$, then the result holds for all p. Moreover, for the Neumann problem, for any $p \le 1$, it does not suffice to assume that the data belongs to $h^p_r(\Omega)$; however, a conclusion analogous to Theorem 2.5.9 is valid if we assume the data to belong to $h^p_z(\Omega)$. In detail:

THEOREM 2.5.9. *Let* $\mathbf{G}$ *be the Green's operator of the Dirichlet problem for the Laplacian on* Ω. *Then the operators*

$$\frac{\partial^2 \mathbf{G}}{\partial x_j \partial x_k}$$

originally defined on $C^\infty(\overline{\Omega})$, *can be extended as bounded operators from* $h^p_z(\Omega)$ *to* $h^p_r(\Omega)$, *for* $1 \le j, k \le N$ *and* $0 < p \le 1$.

THEOREM 2.5.10. *Let* $\tilde{\mathbf{G}}$ *be the Green's operator of the Neumann problem for the Laplacian on* Ω. *Then the operators*

$$\frac{\partial^2 \tilde{\mathbf{G}}}{\partial x_j \partial x_k}$$

originally defined on $C^\infty(\overline{\Omega})$, *can be extended as bounded operators from* $h^p_z(\Omega)$ *to* $h^p_r(\Omega)$, *for* $1 \le j, k \le N$ *and* $0 < p \le 1$.

Some Technical Ideas

We continue to work on a smoothly bounded domain Ω.

PROPOSITION 2.5.11. *Let $0 < p \leq 1$. A function f is in $h_r^p(\Omega)$ if and only if*

$$f = \sum_j \lambda_j a_j,$$

where the a_j are atoms supported on cubes $Q_j \subseteq \Omega$, $\sum_j |\lambda_j|^p < \infty$, and the atoms are of three types:

(1) Atoms a_j that are supported in a "big" cube Q_j (the diameter of Q_j is greater than 1) and satisfy the standard size but not necessarily the moment conditions.

(2) Atoms a_j that are supported in a "small" cube Q_j (the diameter of Q_j is less than or equal to 1) that is "far" from the boundary (distance at least four times the diameter of Q_j) and that satisfy the standard size and cancellation conditions.

(3) Atoms a_j that are supported in a "small" cube Q_j that is "near" to the boundary (distance less than or equal to four times the diameter of Q_j) satisfy the standard size but not necessarily *the moment conditions.*

A result of this nature appears in [**MIY**]. We call a_j a type (a) atom if it satisfies property (1) or (2) and a type (b) atom if it satisfies property (3).

PROPOSITION 2.5.12. *Let $0 < p \leq 1$. A function f is in $h_z^p(\Omega)$ if and only if*

$$f = \sum_j \lambda_j a_j,$$

where the a_j are local p- atoms supported on cubes $Q_j \subseteq \Omega$, $\sum_j |\lambda_j|^p < \infty$, and all the atoms except one are classical atoms satisfying both the size and moment conditions. The one exceptional atom has no cancellation and it can be taken to be supported in a unit cube.

What is interesting in these propositions is the problem of generating an atomic decomposition *with atoms supported entirely in the domain* Ω. Our tool for generating such an atomic decomposition is a version of the classical Calderón reproducing formula. We also use the technique of [**CF**] for proving the atomic decomposition in $H^p(\mathbb{R}^N)$. Thus the area integral is involved.

Now we describe the reproducing formula. There exists a pair of functions Φ and Ψ in $C_c^\infty(\mathbb{R}^N)$, so that a fixed number of moments of Φ and Ψ vanish, and with

$$\Phi_t(x) = t^{-N}\Phi\left(\frac{x}{t}\right), \qquad \Psi_t(x) = t^{-N}\Psi\left(\frac{x}{t}\right)$$

we have

$$f = \int_0^\infty f * \Phi_t(x) * \Psi_t(x)\,\frac{dt}{t}.$$

To construct such a Φ and Ψ, we need only take a radial Φ, whose Fourier transform vanishes sufficiently rapidly at the origin, but with Φ not identically zero. Then we can choose Ψ to be a constant multiple of Φ. The "area integral" that we need is this:

DEFINITION 2.5.13. Let Φ be as in the theorem. Set

$$S_\Phi(f)(x) = \left(\int_{|x-y|<t} |f * \Phi_t(x-y)|^2 \frac{dy\,dt}{t^{N+1}} \right)^{\frac{1}{2}}.$$

THEOREM 2.5.14. *With Φ, S_Φ as above,*

$$\|S_\Phi(f)\|_{L^p} \leq A\|f\|_{H^p}, \qquad \textit{for} \quad 0 < p < \infty.$$

We shall generalize Calderón's formula to a Lipschitz domain in $\mathbb{R}^N$. We now assume that Ω is a *special Lipschitz domain*—that is, the supergraph of a Lipschitz function of $N-1$ variables. The critical geometric property of such a domain Ω is that there is a cone Γ such that, for any $x \in \Omega$, $x + \Gamma$ lies in Ω. The result for a general smoothly bounded domain is then obtained by a simple amalgamation procedure. Now our main result is:

THEOREM 2.5.15. *Fix any open cone $\Gamma \subset \mathbb{R}^N$ and any positive number M. Then there exists $2N$ functions*

$$\Phi^1, \ldots \Phi^N \quad , \quad \Psi^1, \ldots, \Psi^N$$

satisfying

$$\int \Phi^\ell \, dx = 0 \qquad \textit{and} \qquad \int \Psi^\ell x^\alpha \, dx = 0$$

for $\ell = 1, \ldots, N$ and all multi-indices α such that $0 \leq |\alpha| \leq M$. The function Φ^ℓ, Ψ^ℓ are all supported in Γ. Most significantly, if $f \in L^1(\mathbb{R}^N)$ then

$$f = \sum_{\ell=1}^N \int_0^\infty f * \Phi_t^\ell * \Psi_t^\ell \frac{dt}{t}.$$

Observe that if f is supported in Ω then, because we have assumed that Ω is a special Lipschitz domain, we know that $(f * \Phi_t^\ell)(x)$ and $(f * \Phi_t^\ell * \Psi_t^\ell)(x)$ are supported in Ω for all $t > 0$. Thus we may decompose the distribution f into an atomic decomposition whose atoms lie in Ω by the device of breaking up (x, t) space into a sort of Whitney decomposition.

The Reflection Function

In the special case that Ω is the upper half space, then we may exploit the parity of the Newtonian potential with respect to boundary reflection to write down the Green's functions for the Dirichlet and Neumann problems.

In the general case (of a smoothly bounded domain Ω) the global geometry is not so simple. But we shall still have a local notion of reflection.

If $\zeta \in \partial\Omega$ and $x \in \Omega$ is near to ζ then we define the mapping $x \mapsto r_\zeta(x)$ to be the standard reflection of x in the tangent hyper-plane to $\partial\Omega$ at ζ.

A second reflection function is given as follows: If $x \in \Omega$, with x sufficiently close to the boundary, then we define $x \mapsto r(x)$ by $r(x) = r_{\zeta_0}(x)$, where $\zeta_0 = \zeta_0(x)$ is *the* point on the boundary which is nearest to x. Notice that

$$\operatorname{dist}(x, \partial\Omega) = \operatorname{dist}(r(x), \partial\Omega).$$

For us a crucial observation is that

$$|r(x) - r_\zeta(x)| \leq C \cdot |\zeta - x|^2,$$

as $x \to \zeta$. For this estimate one needs C^2 boundary. As we shall see in a moment, the reflection functions are our device for writing down approximate Green's operators for the two boundary value problems that we wish to study.

The Calculation for the Dirichlet Problem

We now sketch the calculation for the Dirichlet problem. Let f be the data function for the Dirichlet problem. We will assume that $f \in h^p_r(\Omega)$, $\frac{N}{N+1} < p \leq 1$. We set

$$f_o(x) = \begin{cases} f(x) & \text{if} \quad x \in \Omega \\ -f\big(r^{-1}(x)\big)\mathcal{J}_{r^{-1}} & \text{if} \quad x \in {}^c(\overline{\Omega}) \end{cases}$$

Here $\mathcal{J}_g$ is the Jacobian of g. The solution to our boundary value problem is then

$$G(f) = E * f_o - P\left(R(E * f_o)\right).$$

Here E is the standard Newtonian potential, R is the restriction operator to the boundary of the domain, and P is the standard Poisson operator for the Laplacian.

Then

$$\frac{\partial^2}{\partial x_i \partial x_j} P\big(R(E * f_o)\big) \equiv \int_\Omega K(x, y) f(y)\, dy,$$

where the kernel K is defined by this equation. Elaborate calculations and the properties of the reflection function described above show that K satisfies the following properties:

$$|K(x, y)| \leq A \cdot |x - y|^{-N+1}$$

and

$$\left|\frac{\partial}{\partial y} K(x, y)\right| \leq A \cdot |x - y|^{-N}.$$

As already noted, in order to solve the Dirichlet problem for *all* $0 < p < \infty$ we must restrict attention to data lying in $h^p_z(\Omega)$. Again, the solution to our boundary value problem is

$$G(f) = E * f_o - P\big(R(E * f_o)\big),$$

but now we have set

$$f_o(x) = \begin{cases} f(x) & \text{if} \quad x \in \Omega \\ 0 & \text{if} \quad x \notin \Omega. \end{cases}$$

The reflection function plays a crucial role once again, for it can be shown that

$$\frac{\partial^2}{\partial x_i \partial x_j} P\left(R(E * f)\right) \equiv \int_\Omega K(x, x - r(y)) f(y)\, dy.$$

The distribution $K(x,z)$ can be seen to have the properties

1: $|\partial_z^\alpha K(x,z)| \leq A_\alpha |z|^{-N-|\alpha|}$;

2: $\widehat{K}(x,\cdot) \in L^\infty$.

These inequalities hold *uniformly* in x and $\partial_x^\beta K(x,z)$ also satisfies these two inequalities (α, β), uniformly in x.

Now we have

LEMMA 2.5.16. *Whenever K satisfies the above properties then the mapping,*

$$f \to \int_\Omega K(x, x - r(y)) f(y) dy,$$

extends to a bounded mapping of $h_z^p(\Omega)$ to $h_r^p(\Omega), 0 < p \leq 1$.

To study the Neumann problem, for all $0 < p < \infty$, we set

$$\tilde{G}(f) = E(f) + P(u_b),$$

where

$$u_b = -(N^+)^{-1} R \frac{\partial E}{\partial \vec{n}}(f).$$

Here N^+ is the "Dirichlet-to-Neumann" operator (see [**CNS1, CNS2**]) Then we are led to study

$$\frac{\partial^2 P}{\partial x_i \partial x_j}\left(N^+\right)^{-1} R \frac{\partial E}{\partial \vec{n}}(f) = \int_\Omega \tilde{K}(x, x - r(y)) f(y)\, dy.$$

It turns out that the kernel $\tilde{K}$ has the same two properties enjoyed by the kernel K that arose in our discussion of the Dirichlet problem. Thus we can show that the operators that comprise $\frac{\partial^2 \tilde{G}}{\partial x_i \partial x_j}$, $i, j = 1, \ldots, N$, originally defined on $C^\infty(\overline{\Omega})$, can be extended as bounded operators from $h_z^p(\Omega)$ to $h_r^p(\Omega)$, $0 < p < \infty$.

We close this discussion by noting that we do not know whether the target spaces of any of these operators can be taken to be h_z^p. The work of Boutet de Monvel ([**BOU1**] - [**BOU3**]) on operators of "transmission type" suggests that the answer may be 'no.'

2.6. Smoothness of Functions. The theory of spaces of homogeneous type in harmonic analysis supports the real variable theory of H^p spaces only for p near 1. A method for extending the range of p is developed in this section. At the same time, intrinsic, coordinate-free methods for considering Hardy spaces and atoms are developed for various contexts.

One upshot of this work is that function spaces that enjoy the functorial properties of smooth functions are defined in contexts where there is no explicit differentiable structure in the ambient space. The reader should keep in mind that some of the material in this section is speculative in nature; it is meant to suggest some approaches to thinking about functions that play the role of smooth objects when the differentiable structure is not presented *a priori*.

Introduction

It is frequently useful in mathematical analysis to discuss smooth functions defined on a set which itself has no explicit smooth (manifold) structure. A classical example is the following question of H. Whitney: when is a function f on a closed set $E \subseteq \mathbb{R}^N$ the restriction to E of a C^k function defined on all of $\mathbb{R}^N$? A rather complete answer—in terms of Taylor series type jets—is given in [**WHI**]. Modern treatments, with additional results, may be found in [**FED**] and [**STE1**]. Whitney was particularly interested in extension results of this sort in the *real analytic* category. As of this writing, that circle of questions has not received a satisfactory treatment.

Put a different way, sometimes it is just too complicated too deal directly with the given smooth structure—especially with the Lie algebra structure. It took many years to see how to handle strongly pseudoconvex domains (see [**FOS1**], [**ROS**]). And it took another several years to see what to do about finite type domains in $\mathbb{C}^2$. That program is still in development. The finite type in $\mathbb{C}^n$ program has barely begun. We hope that some of the ideas presented here might suggest some soft methods for dealing with problems such as this.

In this section we offer a different approach to smooth functions. The idea is to isolate the functorial properties of spaces of Lipschitz functions. Our idea is both motivated by, and bound up with, considerations of real variable Hardy spaces.

Because we are going to be juggling various versions of the axioms for a space of homogeneous type in this section, we indulge ourselves now in a repetition of those axioms for ease of reference:

DEFINITION 2.6.1. A triple (X, d, μ), where (X, μ) is a measure space and $d : X \times X \to \mathbb{R}^+$ is a quasimetric is called a *space of homogeneous type* if

- **a:** If $0 < r < \infty, x \in X$, then the metric ball $B(x, r)$ has positive, finite measure.
- **b:** There is a constant $c > 0$ such that if $B(x, r) \cap B(y, s) \neq \emptyset$ and if $s \geq r$ then $B(y, cs) \supseteq B(x, r)$.
- **c:** There is a constant $c' > 0$ such that $\mu\big(B(x, 2r)\big) \leq c'\mu\big(B(x, r)\big)$ for any $x \in X$ and $r > 0$.

Further recall the definition of atom:

DEFINITION 2.6.2. Let (X, μ) be a space of homogeneous type. A 1-atom is a measurable function $a(x)$ such that

- **i:** a is supported in a metric ball $B(P, r)$;
- **ii:** $|a(x)| \leq 1/\big[\mu(B(P, r)\big]$;
- **iii:** $\int_{B(P,r)} a(x)\, d\mu(x) = 0$.

In fact Coifman and Weiss [**COW2**] showed that there will exist a $p_0 < 1$ (depending on the particular space of homogeneous type being considered) such that atoms defined by (i), (iii), and

$$\textbf{(ii)}_p \qquad |a(x)| \leq \frac{1}{\left[\mu(B(P,r))\right]^{1/p}}$$

give a satisfactory real variable H^p space theory for $p_0 < p \leq 1$.

It has remained an open problem to consider the correct analogy of condition (iii) in the last definition for small p. The problem is that, classically (on $\mathbb{R}^N$), the moment condition for an atom when p is small is

$$(\textbf{iii}_p) \qquad \int_B a(x)x^\alpha\, dx = 0 \qquad \forall |\alpha| \leq \left[N(1/p-1)\right].$$

We call an atom satisfying this explicit moment condition a "classical p-atom."

Even in familiar settings such as the boundary of the unit ball in $\mathbb{C}^2$, it is awkward to formulate the analogue of $(\textbf{iii}_p)$ (see [**GAL**] for some of the complications that arise).

The thrust of the present discussion is to address methods for generalizing $(\textbf{iii}_p)$. We will begin by giving a coordinate-free rendition of $(\textbf{iii}_p)$ that allows us to define real-variable Hardy spaces, for the full range of p, on manifolds; in particular, this construct is valid on the boundary of a domain in $\mathbb{R}^N$. We will then indicate how this construct can be adapted to the boundary of various types of domains in $\mathbb{C}^n$ and to stratified nilpotent groups.

Next we will borrow an idea of Coifman and Weiss ([**WEI**]) that gives yet another approach to $(\textbf{iii}_p)$ that can be adapted to spaces of homogeneous type. We will be able to define real Hardy spaces, for the full range of $0 < p \leq 1$, on a space of homogeneous type.

Finally, we use the classical fact that the dual of the Hardy space $H^p(\mathbb{R}^N)$ is the Lipschitz space $\dot{\Lambda}_\alpha$ $(\mathbb{R}^N)$, $\alpha = N(1/p - 1)$ (see [**FST**]), to define Lipschitz spaces of order $\alpha > 1$ on a space of homogeneous type (heretofore the theory was limited to what can be done in a completely arbitrary metric space—namely Λ_α for $0 < \alpha < 1$ can be defined and classical Lip_1 can be defined). Note that $\dot{\Lambda}_\alpha$, as opposed to Λ_α, denotes the Lipschitz spaces with a smoothness condition imposed but no growth at infinity. Details of this distinction will be explicated below.

We thank Michael Frazier and Guido Weiss for helpful discussions.

Moment Conditions in a Coordinate-Free Context

Our view here is to recast what is already known in the classical setting in Euclidean space, with a view to generalizing to other contexts. The first step is to free the moment condition from coordinates. We work on $\mathbb{R}^N = \partial\mathbb{R}^{N+1}$. Fix $k \in \{0, 1, 2, \dots\}$ and $N/(N + k + 1) < p \leq N/(N + k)$. We will define a "generalized p atom" as in the introduction, except that (iii) (resp. $(\textbf{iii}_p)$) is replaced by the new condition

$$\left|\int a(x)\phi(x)\,d\right| \leq \|\phi\|_{C^{k+1}(\overline{B(0,r)})} \cdot r^{N+k+1-N/p}$$
$$\text{for all } \phi \in C_c^\infty(B(0,2r)). \qquad (\mathbf{iii}_p^*)$$

Remark: It is immediate that any classical H^p function has an atomic decomposition using the new atoms defined with ($\mathbf{iii}_p^*$) just because the classical atoms (defined with (iii)) are also generalized atoms.

However it is not the case that every generalized atom is a classical atom. This so because every sufficiently small perturbation of a generalized p-atom is also a generalized p-atom. ∎

PROPOSITION 2.6.3. *Let*

$$K(z) = \frac{\Omega(x)}{|x|^N}$$

be a standard Calderón-Zygmund convolution kernel, with Ω homogeneous of degree zero and C^∞ smooth away from the origin. Then the operator

$$T\eta(x) = P.V. \int K(x-t)\eta(t)\,dt,$$

initially defined for $\eta \in C_c^\infty(\mathbb{R}^N)$, extends to an operator that maps generalized p-atoms into $L^p, 0 < p \leq 1$. The bound here depends on p but not on the particular atom.

COROLLARY 2.6.4. *The operator in the proposition maps $H^p(\mathbb{R}^N)$ into $H^p(\mathbb{R}^N)$, $0 < p \leq 1$.*

PROOF OF THE COROLLARY: Any function in $H^p(\mathbb{R}^N)$ has a decomposition into classical atoms (see [**LAT**]), and classical atoms are special instances of generalized atoms. Since, by the proposition, generalized atoms are mapped into L^p with a uniform bound, we find that the operator sends H^p into L^p. Now an application of the Riesz transforms, as in [**FST**], finishes the proof of the corollary. ∎

PROOF OF THE PROPOSITION: The case $1 < p < \infty$ being classical and of no interest, we fix $0 < p \leq 1$. Let $k \in \{0, 1, 2, \dots\}$ be chosen so that

$$\frac{N}{N+k+1} < p \leq \frac{N}{N+k}.$$

Fix a generalized p-atom a. Without loss of generality, we may take a to be supported in a ball $B(0,r)$ centered at the origin. Mimicking the classical theory (see [**STE1**]), we estimate $\|a * K\|_{L^p}$ in two pieces: (**a**) The case $|x| \leq 3r$:

$$\begin{aligned}\int_{|x|\leq 3r} |a * K(x)|^p\,dx &= \int_{|x|\leq 3r} |a*K(x)|^2\,dx^{p/2} \cdot \int_{|x|\leq 3r} 1^{2/(2-p)}\,dx^{(2-p)/2} \\ &\leq \|a*K\|_{L^2}^p \cdot C \cdot \left(r^N\right)^{(2-p)/2} \\ &\leq C'\|a\|_{L^2}^p r^{N(2-p)/2} \\ &\leq C''\left(r^{N/2-N/p}\right)^p r^{N-Np/2} \\ &= C''.\end{aligned}$$

In the penultimate inequality we have used the fact that Calderón-Zygmund operators are bounded on L^2. In the last inequality we have used properties (i) and (ii) of an atom to estimate the size of a. (**b**) The case $|x| > 3r$: Let ψ be a cutoff function satisfying

i: $\psi(t) \equiv 1$ for $|t| \leq r$;

ii: $\psi(t) \equiv 0$ for $|t| > 3r/2$.

Then

$$\int_{|x|>3r} |a*K(x)|^p\,dx = \int_{|x|>3r} \left|\int_{|t|\leq r} a(t)\{\psi(t)K(x-t)\}\,dt\right|^p dx. \tag{1}$$

Observe that, for fixed x, the function

$$t \mapsto \psi(t)K(x-t)$$

lies in $C_c^\infty(0, 2r)$. So we may estimate the inside integral in the last expression by using condition ($\mathbf{iii}_p^*$). Thus the right side of (1) is majorized by

$$\begin{aligned}&\int_{|x|>3r} \left\{\|\psi(\cdot)K(x-\ \cdot\ )\|_{C^{k+1}(\overline{B(0,r)})} \cdot r^{N+k+1-N/p}\right\}^p dx \\ &\leq \int_{|x|>3r} \left\{|x|^{-N-k-1} r^{N+k+1-N/p}\right\}^p dx \\ &\leq C \cdot \left(r^{Np+kp+p-N}\right) \cdot r^{-Np-kp-p+N}.\end{aligned}$$

Notice that the last estimation is correct (the integral converges) because $-Np - kp - p + N < 0$. Of course the last line is majorized by a universal constant.

We conclude that the operator maps H^p to L^p. The proof is therefore complete. ∎

Remarks: (**a**) In contexts for which there is no theory of Riesz transforms, more complicated arguments in the spirit of those just presented may be used to prove that

$$\|(Ta)^*\|_{L^p} \leq C \cdot \|a\|_{H^p}, \tag{2}$$

where * is a *truncated* maximal operator as used in [**GOL**]. That is, it is analogous to operators $*$ defined elsewhere in this monograph, but the supremum is taken only over $0 < t < 1$. The arguments for proving (2) are given in [**KRA13**], [**CKS2**].

(b) It is sometimes useful to note that the moment condition that works for H^p when $N/(N+k+1) < p \le N/(N+k)$ also works for $p > N/(N+k)$. The second part of the proof of the proposition makes this clear. One simply needs a sufficient order of decay at infinity so that the integral converges. This observation will be useful, for instance, in the context of doing analysis on the boundary of a finite type domain in $\mathbb{C}^n$. One may tailor atoms for the entire boundary by using the moment condition for points of the highest type that occurs.
(c) The ideas presented so far apply rather naturally to stratified nilpotent lies groups (see [**FOL2**]) and to the "geometry generated by vector fields" situations (see [**NSW2**]). More explicitly, we define an atom in this context to be a measurable function supported in a ball B (where the ball is defined in the relevant geometry) and with (ii) replaced by

$$|a(x)| \le \big(\mathrm{vol}(B)\big)^{-1/p}$$

and condition (iii) replaced by

$$\Big|\int a(t)\phi(t)\,dt\Big| \le \|\phi\|_{\mathcal{C}^{k+1}(\bar{B})} \cdot r^{\mathcal{N}+k+1-\mathcal{N}/p}.$$

Here $\mathcal{C}^\ell$ is the usual non-isotropic space with norm

$$\|\phi\|_{\mathcal{C}^\ell} \equiv \sum_{w_1(j_1)+\cdots+w_n(j_n)\le\ell} \sup |X_1^{j_1}\cdots X_m^{j_m}\phi|,$$

where we are counting multiplicities with weights s. Also, in this last equation, $\mathcal{N}$ denotes the "homogeneous dimension" (see [**FOL2**] for more on this concept).

We shall provide no more details about these variants at this time, but shall only note that it is now easy to see how to define atoms on the boundary of the unit ball in $\mathbb{C}^n$, or more generally on the boundary of a strongly pseudoconvex domains, using the balls of A. Koranyi [**KOR1**] or Stein [**STE3**].

The methods that have been presented in this subsection are used decisively in [**KL2**] to study Hardy spaces in $\mathbb{C}^n$.

The Mean Value Property by Way of Canonical Kernels

In the present subsection we borrow on an idea of Coifman and Weiss ([**WEI**]) to develop a method of expressing higher order mean value conditions without reference, either directly or indirectly, to balls, to monomials, or to Taylor expansions.

Thus our aim is to present a new notion of p-atom, which we will call a "homogeneous p-atom", beginning with the most classical setting of $\mathbb{R}^N$.

DEFINITION 2.6.5. A measurable function a on $\mathbb{R}^N$ is called a *homogeneous p-atom* if it satisfies the following three conditions:

i$_h$**:** a is supported in a Euclidean ball $B = B(x,r)$.

ii$_h$**:** $|a(x)| \le [m(B)]^{-1/p}$, where m is Euclidean volume measure.

iii$_h$: The function

$$A(x) = \int \frac{a(t)}{1+|x-t|^N}\, dt$$

satisfies $A \in L^p(\mathbb{R}^N)$.

Now the first main result of this section is:

PROPOSITION 2.6.6. *A homogeneous p-atom is in fact a classical p-atom.*

PROOF. The proof resembles in spirit the classical Riesz theory of homogeneous distributions (see [**HOR4**]).

First consider the case $N/(N+1) < p \le 1$. If a does not satisfy the classical mean value property, say that $\int a(x)\, dx = c \ne 0$, then $A(x) \approx 1/(1+|x|^N)$ for x large. But then it plainly follows that $A \notin L^p$ for x near ∞, any $p \le 1$. So a does not satisfy (**iii$_h$**).

Conversely, if a is a classical p-atom, hence satisfies the classical moment condition, then for $|x| > 3r$ we have

$$\begin{aligned} |A(x)| &\equiv \left| \int \frac{a(t)}{1+|x-t|^N}\, dt \right| \\ &= \left| \int \frac{a(t)}{1+|x-t|^N}\, dt - \int \frac{a(t)}{1+|x|^N}\, dt \right| \\ &\le C \cdot \int \frac{|a(t)| \cdot |t|}{1+|x-t|^{N+1}}\, dt \\ &\le C \cdot \frac{1}{|x|^{N+1}}. \end{aligned} \tag{3}$$

This last is in L^p for $p > N/(N+1)$. Take care to notice that the size and support of a are of no interest in this last calculation (only that a is bounded and has compact support).

The argument for higher moments is similar: if

$$\frac{N}{N+k+1} < p \le \frac{N}{N+k}$$

and a does not satisfy the corresponding classical moment condition, then let $\ell \le k$ be the order of the first moment of a that does not vanish. It follows from subtracting off a Taylor expansion of the kernel $1/(1+|x-t|^N)$ that $|A(x)| \ge C/|x|^{N+\ell}$ when x is large. As a result, A is not p^{th} power integrable at ∞. Conversely, if a does satisfy the classical moment condition then we may subtract off a higher Taylor polynomial in step (3) above and see that $A \in L^p$.

This completes the proof of the proposition. ∎

Of course the reader will recognize that the kernel $M(x,t) \equiv 1/(1+|x-t|^N)$ is essentially the (undilated) Poisson kernel for Euclidean space. It is natural to formulate moment conditions on a space of homogeneous type by using an analogous kernel. Therefore we must isolate the crucial geometric properties of this kernel. We begin first with *non-compact* spaces of homogeneous type.

In a context where smoothness makes sense, it would be natural to isolate the essential properties of M as follows:

1: M and each of its derivatives in t is bounded.

2: For fixed x, $|M(x,t)| \leq C \cdot m(B(x, |t-x|))^{-1}$.

3: For fixed x, $|\partial_t^\alpha M(x,t)| \leq C \cdot [m(B(x, |t-x|))]^{-(N+|\alpha|)/N}$. .

We leave it for the interested reader to check that any M satisfying these three properties can be used to characterize the mean value properties as we have done in the last proposition.

The discussion in the last paragraph is attractive from the point of view of spaces of homogeneous type because it lends itself readily to the language of *measure distance* ρ: In the language of measure distance, properties (2) and (3) above for the kernel M above may be expressed as

2': For fixed x, $|M(x,t)| \leq C \cdot [\rho(x,t)]^{-1}$.

3': For fixed x, $|\partial_t^\alpha M(x,t)| \leq C \cdot [\rho(x,t)]^{-(N+|\alpha|)/N}$.

However, for a strictly general notion of space of homogeneous type we may not have a notion of dimension (played here by N) so it is more convenient to express (3') as **(3*)** For fixed x,

$$\left| \frac{\partial}{\partial t^\alpha} M(x,t) \right| \leq C \cdot \frac{[d(x,t)]^{-|\alpha|}}{\rho(x,t)}.$$

Again, one checks easily that in the classical setting the proof of the proposition goes through for any kernel M satisfying (1), (2'), (3') or (1), (2'), (3*).

But of course our goal is to transplant the theory to settings where the differentiable structure is either not given or not apparent, so we must specify (3) (resp. (3'), (3*)) rather differently. We turn to this task in the next section.

Hardy Spaces Defined with Respect to a Ponderance

We use here the word "ponderance" in part because the word "weight" already has a well-established (and different) use in harmonic analysis and in part because in nuance this new word more closely suits our purposes.

DEFINITION 2.6.7. Let (X, μ) be a space of homogeneous type. Let $M(x,y)$ be a jointly measurable function on $X \times X$ which is bounded on compact subsets of $X \times X$. Let $0 < p \leq 1$. We say that a compactly supported and measurable function α on X satisfies the *moment condition of order p with respect to the ponderance M* if the function

$$A(x) \equiv \int_X \alpha(t) M(x,t)\, dt$$

is p^{th} power integrable on X.

DEFINITION 2.6.8. Let (X, μ) be a space of homogeneous type. Let $M(x,y)$ be a jointly measurable function on $X \times X$ which is bounded on compact subsets of $X \times X$. Let $0 < p \leq 1$. A measurable function a on X is called a p-atom relative to M if it satisfies the following conditions:

i: a is supported in a metric ball $B(P, r)$;

ii: $|a(x)| \leq 1/\left[\mu(B(P,r)\right]^{1/p}$;

iii: a satisfies the moment condition of order p with respect to M.

At this step we have a choice to make. It is convenient to postulate that we have distinguished a space $\mathcal{D}$ of bounded, measurable functions on X with compact support. We will think of these as testing functions. If X is a perfectly arbitrary space of homogeneous type, then of course there is no (given) smooth structure. The space of testing functions might be chosen (as in Section 2.3) to be a space of Lipschitz functions $(\Lambda_\alpha, 0 < \alpha \leq 1)$. If X is, say, the boundary of a smoothly bounded domain (the case of greatest interest to us) then $\mathcal{D}$ could be simply the space of C^∞ functions.

DEFINITION 2.6.9. Let (X, μ) be a space of homogeneous type. Let $M(x, y)$ be a jointly measurable function on $X \times X$ which is bounded on compact subsets of $X \times X$. Let $0 < p \leq 1$. Let $\mathcal{H}^p_{M,0}$ be the space consisting of all finite sums

$$\sum \lambda_j a_j,$$

where the a_j's are p-atoms and the λ_j's are complex constants.

Assume that we have a space $\mathcal{D}$ of testing functions. Let H^p_M be the weak-$*$ closure of $H^p_{M,0}$, thought of as functionals on $\mathcal{D}$ with the usual L^2 pairing.

We call H^p_M the pth order Hardy space with respect to the ponderance M.

Remark: It is possible that one could avoid pre-specifying the space $\mathcal{D}$ of testing functions by use of more functional analysis. We defer that consideration to another time. [Refer to [**KRA15**] for another view of spaces of testing functions in a related context.]

EXAMPLE 2.6.10. As we have explained in earlier sections, the classical real variable Hardy spaces on $\mathbb{R}^N$ equipped with the usual metric and measure are just those modeled on the ponderance coming from the Poisson kernel. Of course our space of testing functions should be just the usual space C^∞_c of smooth functions of compact support. ∎

EXAMPLE 2.6.11. Let the space of homogeneous type be Euclidean space, as in the last example. Let $\phi \in C^\infty_c(\mathbb{R}^N)$. Define $M(x, y) = \phi(x - y)$. If $\alpha(x)$ is compactly supported and measurable then the moment condition of order p on α is trivially satisfied, all $0 < p \leq 1$. Let the testing functions be as in Example 2.6.10.

In this case the p-atoms are essentially multiples of characteristic functions and the Hardy spaces with respect to the ponderance M are just the usual L^p spaces. ∎

EXAMPLE 2.6.12. Let $\mathbb{H}_n$ be the Heisenberg group with its Haar measure the usual Lebesgue measure on $\mathbb{R}^{2n+1}$ Let τ be the usual non-isotropic distance on $\mathbb{H}_n$. Equipped with this metric and measure, the Heisenberg group forms a space of homogeneous type. (See [**BCK**] for more on the basic ideas here.) Let

$$M(z, \zeta) = \frac{1}{1 + [\tau(z, \zeta)]^{n+1}}.$$

Here $n+1$ is the standard homogeneous dimension of the Heisenberg group. As a manifold, the Heisenberg group is just Euclidean space. So we let the testing functions be as in the first two examples.

Then one may see, although it requires some calculations (see [**FOS2**] for details) that the Hardy spaces H^p_M are the standard real variable Hardy spaces on $\mathbb{H}_n$.

We note in passing (as this will be of interest below) that the ponderance M could have been constructed in more canonical fashion from the Szegö or Poisson-Szegö kernels of the Siegel upper half space that $\mathbb{H}_n$ bounds in a natural way (again consult [**BCK**]). ■

EXAMPLE 2.6.13. Take the space of homogeneous type to the be $\mathbb{Z}$ equipped with counting measure and the usual Euclidean notion of distance. Let

$$M(x,t) = \frac{1}{1+|x-t|}.$$

For simplicity restrict attention to $p=1$.

For the testing functions on this space there is only one choice: the functions supported at finitely many points.

One sees immediately that there is no atom supported at a single point and that every atom must have integral zero. It is not difficult to see that H^1_M consists of all sequences on $\mathbb{Z}$ that sum conditionally. ■

Refer back to the second half of the proof of Proposition 2.6.3. Notice that the definition of H^p_M is designed to make that part (the "hard part") of the proof of the proposition automatic *provided that the ponderance M is qualitatively like the integral kernel being studied.* For instance, a classical Calderón-Zygmund kernel $\Omega(x)/|x|^N$ is qualitatively like the kernel $1/(1+|x|^N)$ because their difference induces a an operator whose kernel acts like a smoothing operator for x large.

In the next subsection we begin to draw all of our ideas together and to develop a notion of "smooth function" on a space of homogeneous type.

The Functorial Properties of a Smooth Function

Let $U \subseteq \mathbb{R}^N$ be a connected open set. We define Lipschitz spaces on U as follows:

For $0<\alpha<1$: Set

$$\Lambda_\alpha(U) = \left\{ f \text{ on } U : \sup_{\substack{x,x+h\in U\\ h\neq 0}} \frac{|f(x+h)-f(x)|}{|h|^\alpha} + \|f\|_{L^\infty} \equiv \|f\|_{\Lambda_\alpha} < \infty \right\}.$$

For $\alpha=1$: Set

$$\Lambda_1(U) = \left\{ f \in C(U) : \sup_{\substack{x,x+h,x-h\in U\\ h\neq 0}} \frac{|f(x+h)+f(x-h)-2f(x)|}{|h|} \right.$$

$$\left. + \|f\|_{L^\infty} \equiv \|f\|_{\Lambda_\alpha} < \infty \right\}.$$

For $\alpha > 1$: Set

$$\Lambda_\alpha(U) = \Big\{ f \text{ on } U : f \in C^1(U) \text{ and}$$

$$\|f\|_{\Lambda_{\alpha-1}(U)} + \sum_{j=1}^{N} \left\| \frac{\partial}{\partial x_j} f \right\|_{\Lambda_{\alpha-1}(U)} \equiv \|f\|_{\Lambda_\alpha} \Big\}.$$

See [**KRA2**] for details on these spaces. The *homogeneous* versions of these spaces are defined in just the same way except that the terms $\|f\|_{L^\infty}$ is omitted from the definitions. The homogeneous spaces are denoted by $\dot{\Lambda}_\alpha$.

The following result is proved in [**DRS**] in dimension 1 and in [**FST**] in higher dimensions:

THEOREM 2.6.14. *Let $0 < p < 1$. The dual of the Frechet space $H^p(\mathbb{R}^N)$ is $\dot{\Lambda}_\alpha$, where $\alpha = N(1/p - 1)$.*

Of course it should be noted in passing that the dual of H^1 is the space of functions of bounded mean oscillation (see [**FST**]). By the Campanato-Morrey theory ([**KRA2**]), BMO is the natural limit, as $\alpha \to 0$, of $\dot{\Lambda}_\alpha$.

The analogous result for the duals of Hardy spaces on the ball in $\mathbb{C}^n$ and related ideas appear [**CRW**], in [**GAL**], and in [**KRA15**].

We should now like to present a proof of this last theorem using the approach to atoms, particularly to the mean value conditions, that was developed in the last subsection. We note in passing that it is rather simple, using the imbedding theorems for Campanato-Morrey spaces (see [**KRA2**]), to verify the pairing between elements of the Hardy class and elements of the corresponding homogeneous Lipschitz class *provided that one uses the classical notion of mean value property.* Using the new more flexible notion (with the weight $M(x,t) = 1/(1 + |x - t|^N)$) requires considerable extra work. But it is the foothold to looking at more general contexts.

PROPOSITION 2.6.15. *We work in $\mathbb{R}^N$. Fix $0 < p \leq 1$. Suppose that f is a locally integrable function that pairs with all p-atoms (where the atoms are defined by (i), (ii), (iii$_h$) from the last section). Then $f \in \dot{\Lambda}_\alpha$, where $\alpha = (1/p - 1) \cdot N$.*

PROOF. For simplicity we consider only the case $0 < \alpha < 1$, that is $N/(N + 1) < p < 1$.

Fix $\phi \in C_c^\infty(\mathbb{R}^N)$ with $\int \phi(x)\, dx = 1$ and ϕ supported in the unit ball. For $s > 0$ we define $\phi_s(x) = s^{-N}\phi(x/s)$. Using a well-known characterization of

Lipschitz spaces (see [**KRA2**]), we consider $(d/ds)(f * \phi_s)$. Observe that

$$\begin{aligned}\left|\int \frac{\frac{d}{ds}\phi_s(t)}{1+|x-t|^N}\,dt\right| &= \left|\frac{d}{ds}\int \frac{\phi_s(t)}{1+|x-t|^N}\,dt\right| \\ &= \left|\frac{d}{ds}\int \frac{\phi(t)}{1+|x-st|^N}\,dt\right| \\ &\leq C\cdot\int \frac{|\phi(t)||t|}{(1+|x-st|)^{N+1}}\,dt.\end{aligned}$$

Therefore, as a function of x, the expression

$$\alpha(x) \equiv \int \frac{\frac{d}{ds}\phi_s(t)}{1+|x-t|^N}\,dt$$

lies in L^p. So $(d/ds)\phi_s$ satisfies the mean value condition $(\mathbf{iii}_h)$.

Further note that

$$\left|\frac{d}{ds}\phi_s\right| \sim s^{-N-1} = s^{-N-1+N/p}\cdot s^{-N/p}.$$

Since ϕ_s and its derivatives are supported in $B(0,s)$, our two calculations taken together show that the function $(d/ds)\phi_s$ equals $C\cdot s^{-N-1+N/p}$ times a generalized p-atom η. Therefore

$$\begin{aligned}\frac{d}{ds}[f*\phi_s] &= f*\left[\frac{d}{ds}\phi_s\right] \\ &= C\cdot s^{-N-1+N/p}\int f(x-t)\eta(t)\,dt.\end{aligned}$$

Our hypotheses about f pairing with H^p now allows us to conclude that

$$\left|\frac{d}{ds}[f*\phi_s]\right| \leq C\cdot s^{-N-1+N/p}.$$

A classical characterization of Lipschitz spaces now tells us that $f \in \dot{\Lambda}_\alpha$ with $\alpha = N(1/p-1)$. That is the desired result. ■

For a converse we have:

PROPOSITION 2.6.16. *Suppose that $f \in \dot{\Lambda}_\alpha, \alpha > 0$. If a is a generalized p-atom, where $\alpha = N\cdot(1/p-1)$, then a pairs with f with a bound depending only on the Lipschitz norm of f.*

The proof of this proposition will take up the remainder of the subsection. It proceeds by way of some Fourier analysis, and is in several steps.

LEMMA 2.6.17. *For appropriate functions f and ϕ it holds that*

$$\begin{aligned}\int_{\mathbb{R}^N} f \cdot \phi \, dx &= \int_{\mathbb{R}^N} \hat{f}(\xi)\hat{\phi}(-\xi)\, d\xi \\ &= c_k \int_{\mathbb{R}^N} \hat{f}(\xi)\hat{\phi}(-\xi)(-2\pi|\xi|)^{2k} \left\{ \int_0^\infty t^{2k-1} e^{-4\pi|\xi|t}\, dt \right\} d\xi\end{aligned}$$

provided that $k \in \{1, 2, \dots\}$ and $c_k > 0$ is chosen appropriately.

PROOF. In fact

$$\begin{aligned}\int_0^\infty t^{2k-1} e^{-4\pi|\xi|t}\, dt &= -\frac{1}{4\pi|\xi|} e^{-4\pi|\xi|t} t^{2k-1} \Big|_0^\infty + \frac{2k-1}{4\pi\xi} \int_0^\infty t^{2k-2} e^{-4\pi|\xi|t}\, dt \\ &= \frac{2k-1}{4\pi\xi} \int_0^\infty t^{2k-2} e^{-4\pi|\xi|t}\, dt \\ &= \cdots \\ &= \frac{(2k-1)(2k-2)\cdots 1}{(4\pi|\xi|)^{2k-1}} \int_0^\infty e^{-4\pi|\xi|t}\, dt \\ &= \frac{(2k-1)!}{(4\pi|\xi|)^{2k}}.\end{aligned}$$

Thus if $c_k > 0$ is selected properly then our assertion becomes true. ■

COROLLARY 2.6.18. *If f, ϕ are suitable functions and $k \in \{1, 2, \dots\}$ then*

$$\int_{\mathbb{R}^N} f(x)\phi(x)\, dx = c_k \int_{\mathbb{R}^{N+1}_+} t^{2k-1} \frac{\partial u}{\partial t^k}(x,t) \frac{\partial v}{\partial t^k}(x,t)\, dt,$$

where u is the Poisson integral of f to the upper half space and v is the Poisson integral of ϕ.

PROOF. Immediate from the lemma and the formula for the Fourier multiplier corresponding to the Poisson kernel. ■

LEMMA 2.6.19. *Let $N/(N+1) < p \le 1$. Suppose that a is a bounded measurable function supported in $B(0,r) \subseteq \mathbb{R}^N$ such that*

$$A(x) = A_0(x) \equiv \int \frac{a(t)}{1 + |x - w|^N}\, dw$$

is a pth power integrable function. Then $A(x)$ is asymptotically of size $\mathcal{O}\big(r \sup |a| \cdot |x|^{-N-1}\big)$ as $|x| \to \infty$.

PROOF. We have already seen that if a does not have mean value zero in the classical sense then the hypothesis of the lemma is not true. Thus $\int a(x)\, dx = 0$ and the result follows immediately from

$$A(x) = \int a(t) \left[\frac{1}{1 + |x - w|^N} - \frac{1}{1 + |x|^N} \right] dw.$$

■

LEMMA 2.6.20. *Let $N/(N+1) < p \leq 1$. Let a be a bounded measurable function supported in $B(0,r) \subseteq \mathbb{R}^N$ such that*

$$A(x) = A_0(x) \equiv \int \frac{a(w)}{1+|x-w|^N}\, dw$$

is a pth power integrable function. Then

$$A_k(x) \equiv \int \frac{a(w)}{1+|x-w|^{N+k}}\, dw$$

is asymptotically of size $\mathcal{O}\big(r \sup|a| \cdot |x|^{-N-k-1}\big)$ as $|x| \to \infty$.

PROOF. Imitate the proof of the last lemma. ∎

LEMMA 2.6.21. *Let a be a bounded measurable function supported in $B(0,r) \subseteq \mathbb{R}^N$ such that*

$$A(x) = A_0(x) \equiv \int \frac{a(w)}{1+|x-w|^N}\, dw$$

is a pth power integrable function. Then, for $t > 0$, the expression

$$\int\int \frac{a(w)}{(t^2+|x-w|^2)^{(N+k)/2}}\, dw dx$$

is of size $\mathcal{O}\big(r^{N+1} \cdot \sup|a| \cdot t^{-k-1}\big)$.

PROOF. As in the previous two lemmas. ∎

PROOF OF THE PROPOSITION: Assume for simplicity that $0 < \alpha < 1$. We will apply the last corollary with $k = 1$, u the Poisson integral of f, and v the Poisson integral of the generalized p-atom a. Let a be supported in a ball B of radius r, centered at the origin without loss of generality, and assume that $|a| \leq r^{-N/p}$.

We calculate that

$$\begin{aligned} \left|\int_{\mathbb{R}^N} f(x)a(x)\, dx\right| &= \left|\int_0^\infty \int_x t\frac{\partial u}{\partial t}\frac{\partial v}{\partial t}\, dxdt\right| \\ &\leq C \cdot \left[\left|\int_{0<t<r} \int_x \, dxdt\right| + \left|\int_{r\leq t<\infty} \int_x \, dxdt\right|\right] \\ &\equiv I + II. \end{aligned}$$

Now

$$\begin{aligned} II &\leq \left| \int_r^\infty \int_x t \cdot \frac{\partial u}{\partial t} \cdot \left[\frac{\partial}{\partial t} \int_w \frac{t \cdot a(w)}{(t^2 + (x-w)^2)^{(N+1)/2}} \, dw \right] \right| dx \, dr \\ &\leq \left| \int_r^\infty \int_x t \cdot \frac{\partial u}{\partial t} \cdot \left[c \cdot \frac{a(w)}{(t^2 + (x-w)^2)^{(N+1)/2}} \right. \right. \\ &\qquad \left. \left. + c' \cdot \frac{t^2 \cdot a(w)}{(t^2 + (x-w)^2)^{(N+3)/2}} \right] \right| dx \, dr \\ &\leq \int_r^\infty t \cdot t^{\alpha - 1} r^{N+1} \cdot r^{-N/p} \cdot t^{-2} \, dr. \end{aligned}$$

Here we have used the standard characterization of Lipschitz functions in terms of their Poisson integrals (see [**STE1**] or [**KRA2**]). This last does not exceed

$$C \cdot \int_r^\infty t^{\alpha-2} r^{N+1-N/p} dr = C,$$

since $\alpha = N(1/p - 1)$. Notice that the integral in r converges because $\alpha - 2 < -1$.

Now we estimate I. We first integrate by parts in t to throw an extra derivative on the expression $t\partial u/\partial t$. Thus I consists of two terms, a typical one of which is

$$\int_0^r \int_x \int_w t \frac{\partial^2 u}{\partial t^2} \left[\int_w \frac{ta(w)}{(t^2 + |x-w|^2)^{(N+1)/2}} \, dt \right] dx dt.$$

Using the Poisson integral characterization of Lipschitz functions, together with the last lemma, we find that this is majorized in absolute value by

$$\int_0^r t^{\alpha-1} r^{N-N/p} dt \leq C,$$

because $\alpha = N(1/p - 1)$. The integral in r converges because $\alpha - 1 > -1$.

We have proved the proposition. ■

Our purpose here has been a matter of form rather than substance. Indeed the classical moment condition lies only slightly below the surface. But we wanted to bring out the role of the Poisson expressions (modeled on the ponderance M) in these duality arguments. This material has been provided for motivation.

Definition of Lipschitz Functions on a Space of Homogeneous Type

Now let (X, d, μ) be a space of homogeneous type and M a ponderance on X. Define spaces H^p_M as above. Motivated by the material in the preceding section, we *define* Lipschitz spaces as follows:

DEFINITION 2.6.22. Let (X, μ), M be as above. Let $0 < p \leq 1$. We define the (homogeneous) Lipschitz space $\dot{\Lambda}_{\alpha,M}$, with $\alpha = 1/p - 1$, to be the Frechet space dual of H^p_M.

Remark: Some remarks are in order. Notice that, in the classical setting of $\mathbb{R}^N$, our correspondence between α and p is off by a factor of N. This is so because of dimensionality considerations. When one defines the measure distance in Euclidean space, it turns out to be the Nth power of ordinary Euclidean distance for similar reasons. One may mediate between the new definition of Lipschitz space given here and the more classical definition of Lipschitz space—by way of the measure distance—using the Campanato-Morrey definition of Lipschitz space. See [**KRA2**] for details.

Boundedness of Integral Operators on Hardy and Lipschitz Spaces

Our Hardy and Lipschitz spaces are designed so that the hard part of the proof of Proposition 2.1 goes through automatically for singular integral kernels which are qualitatively like the ponderance M. In particular, in contexts where the concept of homogeneous dimension is well-defined then Calderòn-Zygmund style singular integrals are also well-defined (see, for instance, [**FOL2**] and [**FOS2**]). The Hardy spaces modeled on a ponderance with asymptotic homogeneity at infinity corresponding to the homogeneous dimension will then be preserved by these singular integral operators. Also the corresponding Lipschitz spaces will be (locally) preserved. Details will not be provided.

Concluding Remarks

Our purpose here has been to present an Ansatz for constructing function spaces in a context with a minimum of structure. The function spaces mesh in a natural way with standard integral operators that arise in harmonic analysis. Examples arise naturally in the context of stratified nilpotent Lie groups, boundaries of domains in $\mathbb{R}^N$ and $\mathbb{C}^n$, and in analysis on manifolds. We hope that this provides some unification of disparate results that have gone before.

We also have provided a coordinate-free method for formulating mean-value properties in a rather general context. This should simplify future analyses of Hardy spaces.

We would be remiss not to mention some other possible methods of considering smooth functions in a rather abstract context. One of these would be the Campanato-Morrey method of characterizing Lipschitz spaces $\dot{\Lambda}_\alpha$ in terms of mean approximation by polynomials. Methods of axiomatizing their approach (in particular of replacing polynomials by families of testing functions that may be constructed on many spaces of homogeneous type) were explored in [**KRA15**].

Finally, if f is a locally integrable, measurable function on a space of homogeneous type, then let $f_{B(x,r)}$ denote the mean of f over the ball $B(x,r)$. Then one might explore the putative smoothness of f by considering differences such as

$$f_{B(x,2r)} - f_{B(x,r)}$$

(this would be a "first difference") and

$$f_{B(x,3r)} + f_{B(x,r)} - 2f_{B(x,2r)}$$

(this would be a "second difference"). Even in the classical Euclidean setting, this turns out to be a surprisingly tricky point of view to develop.

References

[ABA1] M. Abate, Horospheres and iterates of holomorphic maps, *Math. Z.* 198(1988), 225-238.

[ABA2] M. Abate, Iteration theory on weakly convex domains, Seminar in Complex Analysis and Geometry 1988, EditEl, Rende, 1988.

[ABA3] M. Abate, *Iteration Theory of Holomorphic Maps on Taut Manifolds*, Mediterranean Press, Italy, 1989.

[AHL1] L. Ahlfors, *Complex Analysis*, 3rd ed., McGraw-Hill, New York, 1979.

[AHL2] L. Ahlfors, An extension of Schwarz's lemma, *Trans. AMS* 43(1938), 359-364.

[AHL3] L. Ahlfors, *Conformal Invariants*, McGraw-Hill, New York, 1973.

[AHG] L. Ahlfors and H. Grunsky, Über die Blochsche Konstante, *Math. Z.* 42(1937), 671-673.

[AIZ1] L. A. Aizenberg, Multidimensional analogues of the Carleman formula with integration over boundary sets of maximal dimension. (Russian), 12-22, 272, *Akad. Nauk. SSSR Sibirsk. Otdel. Inst. Fiz., Krasnoyarsk*, 1985.

[AIZ2] L. A. Aizenberg, An application of multidimensional Carleman formulas. (Russian. Serbo-Croation summary) Second international symposium on complex analysis and applications (Budva, 1986). *Mat. Vesnik* 38 (1986), 365-374.

[ALA1] G. Aladro, thesis, Penn State University, 1985.

[ALA2] G. Aladro, Applications of the Kobayashi metric to normal functions of several complex variables, *Utilitas Math.* 31(1987), 13-24.

[ALAK] G. Aladro and S. G. Krantz, A criterion for normality in $\mathbb{C}^n$, *Jour. Math. Anal. and Applic.* 161(1991), 1991.

[ALE] H. Alexander, Holomorphic mappings from the ball and polydisc, *Math. Annalen* 209(1974), 249-256.

[ACP] M. Anderson, Clunie, and C. Pommerenke, On Bloch functions and normal functions, *J. Reine Angew. Math.* 270(1974), 12-37.

[AF1] J. Arazy and S. Fisher, Some aspects of the minimal Möbius invariant space of analytic functions on the unit disc, *Lecture Notes in Math.* 1070, Springer, Berlin, 1984.

[AF2] J. Arazy and S. Fisher, The uniqueness of the Dirichlet space among Möbius invariant spaces, *Illinois J. Math.* 29(1985), 449-462.

[AF3] J. Arazy and S. Fisher, Invariant Hilbert spaces of holomorphic functions on bounded symmetric domains, preprint.

[AFP] J. Arazy, S. Fisher, and J. Peetre, Möbius invariant function spaces, *J. Reine Angew. Math.* 363(1985), 110-145.

[BJT] S. Baouendi, H. Jacobowitz, and F. Treves, On the analyticity of CR mappings, *Ann. Math.* 122(1985), 365-400.

[BARK] S. Ross Barker, Two theorems on boundary values of analytic functions, *Proc. A.M.S.* 68(1978), 48-54.

[BAR] D. Barrett, Regularity of the Bergman projection on domains with transverse symmetries, *Math. Ann.* 258(1982), 441-446.

[BEF] R. Beals and C. Fefferman, On local solvability of linear partial differential equations, *Ann. Math.* 97(1973), 482-498.

[BEA] F. Beatrous, L^p estimates for extensions of holomorphic functions, *Mich. Jour. Math.* 32(1985), 361-380.

[BED1] E. Bedford, On the automorphism group of a Stein manifold, *Math. Ann.* 266(1983), 215-227.

[BED2] E. Bedford, Action of the automorphism group of a smooth domain in $\mathbb{C}^n$, *Proc. Am. Math. Soc.* 93(1985), 232-234.

[BEDD] E. Bedford and J. Dadok, Bounded domains with prescribed group of automorphisms, *Comment. Math. Helv.* 62(1987), 561-572.

[BEF] E. Bedford and J. E. Fornæss, A construction of peak functions on weakly pseudoconvex domains, *Ann. Math.* 107(1978), 555-568.

[BEP1] E. Bedford and S. Pinchuk, Domains in $\mathbb{C}^2$ with noncompact group of automorphisms, *Math. Sb.* Nov. Ser. 135(1988), No. 2, 147-157.

[BEP2] E. Bedford, and S. Pinchuk, Domains in $\mathbb{C}^{n+1}$, with non-compact automorphism group, *Journal of Geometric Analysis* 1(1991), 165-192.

[BEP3] E. Bedford, and S. Pinchuk, Convex domains with non-compact automorphism group, preprint.

[BEK1] D. Bekolle, Le dual de l'espace de Bergman A^1 dans la boule de Lie de $\mathbb{C}^n$, *C. R. Acad. Sci. Paris* Series I, Math. 311(1990), 235-238.

[BEK2] D. Bekolle, The Bloch spae and BMO analytic functions in the tube over the spherical cone, *Proc. A.M.S.* 102(1988), 949-956.

[BEK3] D. Bekolle, The dual of the Bergman space A^1 in the tube over the spherical cone, *Recent Progress in Fourier Analysis*, North Holland Studies v. 111, North Holland, Amsterdam, 1985.

[BEBCZ] D. Bekolle, C. A. Berger, L. A. Coburn, K. H. Zhu, BMO in the Bergman metric on bounded symmetric domains, *J. Funct. Anal.* 93(1990), 310-350.

[BEL1] S. Bell, Biholomorphic mappings and the $\bar{\partial}$ problem, *Ann. Math.*, 114(1981), 103-113.

[BEL2] S. Bell, Compactness of families of holomorphic mappings up to the boundary, in *Complex Analysis Seminar*, Springer Lecture Notes vol. 1268, Springer Verlag, Berlin, 1987.

[BEL3] S. Bell, Analytic hypoellipticity of the $\bar{\partial}$- Neumann problem and extendability of holomorphic mappings, *Acta Math.* 147(1981), 109-116.

[BEB] S. Bell and H. Boas, Regularity of the Bergman projection in weakly pseudoconvex domains, *Math. Annalen* 257(1981), 23-30.

[BEK] S. Bell and S. G. Krantz, Smoothness to the boundary of conformal maps, *Rocky Mt. Jour. Math.* 17(1987), 23-40.

[BEEL] S. Bell and E. Ligocka, A simplification and extension of Fefferman's theorem on biholomorphic mappings, *Invent. Math.* 57(1980), 283-289.

[BCK] C. Berenstein, D. C. Chang, and S. G. Krantz, *Analysis on the Heisenberg Group*, to appear.

[BER1] S. Bergman, *The Kernel Function and Conformal Mapping*, Am. Math. Soc., Providence, 1970.

[BER2] S. Bergman, Über die Existenz von Repraäsentantenbereichen, *Math. Ann.* 102(1929), 430-446.

[BFKKMP] B. Blank, Fan Dashan, David Klein, Steven G. Krantz, Daowei Ma, Myung-Myull Pang, The Kobayashi metric of a complex ellipsoid in $\mathbb{C}^2$, *Experimental Math.*, 1(1992), 47-55.

[BLG] T. Bloom and I. Graham, A geometric characterization of points of type m on real submanifolds of $\mathbb{C}^n$, *J. Diff. Geom.* 12(1977), 171-182.

[BOA] H. Boas, Counterexample to the Lu Qi-Keng conjecture, *Proc. Am. Math. Soc.* 97(1986), 374-375.

[BOS] H. Boas and E. Straube, On equality of line type and variety type of real hypersurfaces in $\mathbb{C}^n$, *Jour. Geom. Anal.* 2(1992), 95-98.

[BONK] M. Bonk, On Bloch's constant, *Proc. AMS* 110(1990), 889-894.

[BMS] L. Boutet de Monvel and J. Sjöstrand, Sur la singularité des noyaux de Bergman et Szegö, *Soc. Mat. de France Asterisque* 34-35(1976), 123-164.

[BOU1] L. Boutet de Monvel, Comportement d'un opérateur pseudo-différentiel sur une variéte à bord, *C.R. Acad. Sci. Paris* 261(1965), 4587-4589.

[BOU2] L. Boutet de Monvel, Comportement d'un opérateur pseudo-différentiel sur une variété à bord, I. La propriété de transmission, *J. d'Analyse Mathematique* 17(1966), 241-253.

[BOU3] L. Boutet de Monvel, Comportement d'un opérateur pseudo-différentiel sur une variété à bord, II. Pseudo-noyaux de Poisson, *J. d'Analyse Mathematique* 17(1966), 255-304.

[BUK] D. Burns and S. G. Krantz, Rigidity of holomorphic mappings and a new Schwarz lemma at the boundary, preprint.

[BSW] D. Burns, S. Shnider, and R. O. Wells, On deformations of strictly pseudoconvex domains, *Invent. Math.* 46(1978), 237-253.

[CAR] L. Carleson, Two remarks on H^1 and BMO, *Advances in Math.* 22(1976), 269-277.

[CAT1] D. Catlin, Necessary conditions for subellipticity of the $\overline{\partial}$-Neumann problem, *Ann. Math.* 117(1983), 147-172.

[CAT2] D. Catlin, Subelliptic estimates for the $\bar{\partial}$-Neumann problem, *Ann. Math.* 126(1987), 131-192.

[CAT3] D. Catlin, Global regularity for the $\overline{\partial}$-Neumann problem, *Proc. Symp. Pure Math.* v. 41 (Y. T. Siu ed.), Am. Math. Soc., Providence, 1984.

[CAT4] D. Catlin, Estimates of invariant metrics on pseudoconvex domains of dimension two, *Math. Z.* 200(1989), 429-466.

[CAT5] D. Catlin, private communication.

[CIR] E. Cirka, The theorems of Lindelöf and Fatou in $\mathbb{C}^n$, *Mat. Sb.* 92(134)(1973), 622-644; *Math. U.S.S.R. Sb.* 21(1973), 619-639.

[CHA] D. C. Chang, Estimates for singular integral operators with mixed type homogeneities in Hardy classes, *Math. Ann.* 287(1990), 303-322.

[CF] S. Y. A. Chang and R. Fefferman, A continuous version of duality of H^1 and BMO, *Ann. Math.* 112(1980), 179-201.

[CKS1] D. C. Chang, S. G. Krantz, and E. M. Stein, H^p theory on a smooth domain in $\mathbb{R}^N$ and applications, Proceedings of a Conference in honor of Walter Rudin, American Mathematical Society, Providence, 1993.

[CKS2] D. C. Chang, S. G. Krantz, and E. M. Stein, H^p theory on a smooth domain in $\mathbb{R}^N$ and elliptic boundary value problems, *Jour. Funct. Anal.*, to appear.

[CNS1] D. C. Chang, A. Nagel, and E. M. Stein, Estimates for the $\overline{\partial}$-Neumann problem in pseudoconvex domains in $\mathbb{C}^2$ of finite type, *Proc. Natl. Acad. Sci. USA* 85(1988), 8771-8774.

[CNS2] D. C. Chang, A. Nagel, and E. M. Stein, Estimates for the $\overline{\partial}$-Neumann problem in pseudoconvex domains of finite type in $\mathbb{C}^2$, *Acta Math.*, to appear.

[CHF] S. Y. Chang and R. Fefferman, A continuous version of duality of H^1 with BMO on the bidisc, *Ann. Math.* 112(1980), 179-201.

[CHM] S. S. Chern and J. Moser, Real hypersurfaces in complex manifolds, *Acta Math.* 133(1974), 219-271.

[CHR1] M. Christ, Lectures on Singular Integral Operators, Conference Board of Mathematical Sciences, American Mathematical Society, Providence, 1990.

[CHR2] M. Christ, Regularity properties of the $\overline{\partial}_b-$ equation on weakly pseudoconvex CR manifolds of dimension 3, *Jour. A. M. S.* 1(1988), 587-646.

[CHR3] M. Christ, On the $\bar{\partial}_b$ equaton and Szegö projection on CR manifolds, *Harmonic Analysis and PDE's* (El Escorial, 1987), Springer Lecture

Notes vol. 1384, Springer Verlag, Berlin, 1989.

[CHR4] M. Christ, Embedding compact three-dimensional CR manifolds of finite type in $\mathbb{C}^n$, *Ann. of Math.* 129(1989), 195-213.

[CHR5] M. Christ, Estimates for fundamental solutions of second-order subelliptic differential operators, *Proc. A.M.S.* 105(1989), 166-172.

[CHG] M. Christ and D. Geller, Singular integral characterizations of Hardy spaces on homogeneous groups, *Duke Math. Jour.* 51(1984), 547-598.

[CIK] J. Cima and S. G. Krantz, A Lindelöf principle and normal functions in several complex variables, *Duke Math. Jour.* 50(1983), 303-328.

[CIR] E. Cirka, The theorems of Lindelöf and Fatou in $\mathbb{C}^n$, *Mat. Sb.* 92(134) 1973, 622-644; *Math. USSR Sb.* 21(1973), 619-639.

[CRW] R. R. Coifman, R. Rochberg, and G. Weiss, Factorization theorems for Hardy spaces in several variables, *Ann. of Math.* 103(1976), 611-635.

[COW1] R. R. Coifman and G. Weiss, *Analyse Harmonique non-commutative sur certains espaces homogenes*, Springer Lecture Notes, vol. 242, Springer Verlag, Berlin, 1971.

[COW2] R. R. Coifman and G. Weiss, Extensions of Hardy spaces and their use in analysis, *Bull. Am. Math. Soc.* 83(1977), 569-643.

[COL] F. Colonna, Bloch and normal functions and their relations, *Rend. Circ. Mat. Palermo* 38(1989), 161-180.

[COT] M. Cotlar, A combinatorial inequlaity and its application to L^2 spaces, *Revista Math. Cuyana* 1(1955), 41-55.

[COU] B. Coupet, lecture, Madison, Wisconsin, 1991.

[DAF] G. Dafni, thesis, Princeton University, 1993.

[DAN1] J. P. D'Angelo, *Several Complex Variables and Geometry*, CRC Press, Boca Raton, 1992.

[DAN2] J. P. D'Angelo, Real hypersurfaces, orders of contact, and applications, *Annals of Math.* 115(1982), 615-637.

[DAN3] J. P. D'Angelo, Intersection theory and the $\bar{\partial}$- Neumann problem, *Proc. Symp. Pure Math.* 41(1984), 51-58.

[DAN4] J. P. D'Angelo, Finite type conditions for real hypersurfaces in $\mathbb{C}^n$, in *Complex Analysis Seminar*, Springer Lecture Notes vol. 1268, Springer Verlag, 1987, 83-102.

[DAN5] J. P. D'Angelo, Iterated commutators and derivatives of the Levi form, in *Complex Analysis Seminar*, Springer Lecture Notes vol. 1268, Springer Verlag, 1987, 103-110.

[DAN6] J. P. D'Angelo, Finite type and the intersection of real and complex subvarieties, *Several Complex Variables and Complex Geometry, Proc. Symp. Pure Math.* 52, Part I, 103-118.

[DAN7] J. P. D'Angelo, A note on the Bergman kernel, *Duke Math. Jour.* 45(1978), 259-265.

[DJS] G. David, J. L. Journé, and S. Semmes, Opérateurs de Calderón-Zygmund, fonctions para-accrètive et interpolation, *Rev. Mat. Iboamericana* 1,4(1985), 1-56.

[DIE] K. Diederich, Das Randverhalten der Bergmanschen Kernfunktion und Metrik in streng pseudo-konvexen Gebieten, *Math. Annalen* 187(1970), 9-36.

[DIF1] K. Diederich and J. E. Fornæss, Pseudoconvex domains: Bounded strictly plurisubharmonic exhaustion functions, *Invent. Math.* 39(1977), 129-141.

[DIF2] K. Diederich and J. E. Fornæss, Pseudoconvex domains with real-analytic boundary, *Annals of Math.* 107(1978), 371-384.

[DIH] K. Diederich and G. Herbort, Local extension of holomorphic L^2 functions with weights, *Complex Analysis* (Wuppertal, 1991), 106-110.

[DRS] P. Duren, B. Romberg, and A. Shields, Linear functionals on H^p spaces with $0 < p < 1$, *J. Reine Angew. Math.* 238(1969), 32-60.

[EBI] D. Ebin, The manifold of Riemannian metrics, *Proc. Symp. Pure Math.* XV, Am. Math. Soc., Providence, 1970.

[EIS] D. Eisenman, *Intrinsic Measures on Complex Manifolds and Holomorphic Mappings*, a Memoir of the American Mathematical Society, Providence, 1970.

[EPM] C. Epstein and R. Melrose, Resolvent of the Laplacian on strongly pseudoconvex domains, *Acta Math.* 167(1991), 1-106.

[EZH] V. Ezhov, Cauchy-Riemann automorphism groups that cannot be projectively realized, *Jour. Geom. Anal.* 2(1992), 417-428.

[FRI] E. Fabes and N. Riviere, Symbolic calculus of kernels with mixed homogeneity, *Singular Integrals*, Proc. Symp. Pure Math., American Mathematical Society, Providence, 1966.

[FAK] J. Faraut and A. Koranyi, Function spaces and reproducing kernels on bounded symmetric domains, *J. Funct. Anal.* 88(1990), 64-89.

[FAKR] H. Farkas and I. Kra, *Riemann Surfaces*, Springer, Berlin, 1979.

[FAT] P. Fatou, Series trigonométriques et seéries de Taylor, *Acta Math.* 30(1906), 335-400.

[FED] H. Federer, *Geometric Measure Theory*, Springer, Berlin, 1969.

[FEF] C. Fefferman, The Bergman kernel and biholomorphic mapping of pseudoconvex domains, *Invent. Math.* 26(1974), 1-65.

[FST] C. Fefferman and E. M. Stein, H^p spaces of several variables, *Acta Math.* 129(1972), 137-193.

[FIS] S. Fisher, *Function Theory on Planar Domains: A Second Course in Complex Analysis*, John Wiley and Sons, New York, 1983.

[FOL1] G. B. Folland, The tangential Cauchy-Riemann complex on spheres, *Trans. A. M. S.* 171(1972), 83-133.

[FOL2] G. B. Folland, Subelliptic estimates and function spaces on nilpotent Lie groups, *Ark. Mat.* 13(1975), 161-207.

[FOLK] G. B. Folland and J. J. Kohn, *The Neumann Problem for the Cauchy-Riemann Complex*, Princeton University Press, Princeton, 1972.

[FOS1] G. B. Folland and E. M. Stein, Estimates for the $\bar{\partial}_b$ complex and analysis on the Heisenberg group, *Comm. Pure Appl. Math.* 27(1974), 429-522.

[FOS2] G. B. Folland and E. M. Stein, *Hardy Spaces on Homogeneous Groups*, Princeton University Press, Princeton, 1982.

[FOK] J. E. Fornæss and S. G. Krantz, unpublished.

[FOM] J. E. Fornæss and J. McNeal, A construction of peak functions on some finite type domains, preprint.

[FOS] J. Fornæss and N. Sibony, Construction of p.s.h. functions on weakly pseudoconvex domains, *Duke Math. Jour.* 58(1989), 633-655.

[FRA1] S. Frankel, Complex geometry of convex domains that cover varieties, *Acta Math.* 163(1989), 109-149.

[FRA2] S. Frankel, Applications of affine geometry to geometric function theory in several complex variables, Part I. Convergent rescalings and intrinsic quasi-isometric structure, *Several Complex Variables and Complex Geometry, Proc. Symp. Pure Math.* 52, Part II, 183-208.

[FRJ] M. Frazier and B. Jawerth, The ϕ- transform and applications to distribution spaces, *Lecture Notes in Math.* 1302(1988), 223-246.

[FTW] M. Frazier, R. Torres, and G. Weiss, The boundedness of Calderón-Zygmund operators on the spaces $\dot{F}_p^{\alpha,q}$, *Revista Matemática Iboamericana* 4(1988), 41-72.

[FU] S. Fu, private communication.

[FUKS] B. A. Fuks, *Special Chapters in the Theory of Analytic Functions of Several Complex Variables*, American Mathematical Society, Providence, 1965.

[GAM] T. W. Gamelin, *Uniform Algebras*, Prentice Hall, Englewood Cliffs, 1969.

[GAR] J. Garnett, *Bounded Analytic Functions*, Academic Press, New York, 1981.

[GAL] J. Garnett and R. Latter, The atomic decomposition for Hardy spaces in several complex variables, *Duke Math. J.* 45(1978), 815-845.

[GEB] N. Gebelt, thesis, UCLA, 1992.

[GEL] D. Geller, Princeton U. Press book. *Analytic Pseudodifferential Operators for the Heisenberg Group and Local Solvability*, Princeton University Press, Princeton, 1990.

[GOL] D. Goldberg, A local version of real Hardy spaces, *Duke Math. J.* 46 (1979), 27-42.

[CRG] C. R. Graham, thesis, Princeton University, 1981.

[GRA1] I. Graham, Boundary behavior of the Caratheéodory and Kobayashi metrics on strongly pseudoconvex domains in $\mathbb{C}^n$, *Trans. Am. Math. Soc.* 207(1975), 219-240.

[GRA2] I. Graham, Boundary behavior of the Carathéothodory and Kobayashi metrics, *Proc. Symposia in Pure Math.*, v. XXX, Part I, American Mathematical Society, Providence, 1977.

[GK1] R. E. Greene and S. G. Krantz, Biholomorphic self-maps of domains, *Complex Analysis II* (C. Berenstein, ed.), Springer Lecture Notes, vol. 1276, 1987, 136-207.

[GK2] R. E. Greene and S. G. Krantz, Stability properties of the Bergman kernel and curvature properties of bounded domains, *Recent Progress in Several Complex Variables*, Princeton University Press, Princeton, 1982.

[GK3] R. E. Greene and S. G. Krantz, Deformation of complex structures, estimates for the $\bar{\partial}$ equation, and stability of the Bergman kernel, *Adv. Math.* 43(1982), 1-86.

[GK4] R. E. Greene and S. G. Krantz, Stability of the Carathéodory and Kobayashi metrics and applications to biholomorphic mappings, *Proc. Symp. in Pure Math.*, Vol. 41 (1984), 77-93.

[GK5] R. E. Greene and S. G. Krantz, The automorphism groups of strongly pseudoconvex domains, *Math. Annalen* 261(1982), 425-446.

[GK6] R. E. Greene and S. G. Krantz, Characterizations of certain weakly pseudo-convex domains with non-compact automorphism groups, in *Complex Analysis Seminar*, Springer Lecture Notes 1268(1987), 121-157.

[GK7] R. E. Greene and S. G. Krantz, Invariants of Bergman geometry and the automorphism groups of domains in $\mathbb{C}^n$, *Proceedings of a Conference on Complex Analysis* held in Cetraro, Italy, Mediterranean Press, 1992.

[GK8] R. E. Greene and S. G. Krantz, Techniques for Studying the Automorphism Groups of Weakly Pseudoconvex Domains, Proceedings of the Special Year at the Mittag-Leffler Institute (J. E. Fornæss and C. O. Kiselman, eds.) *Annals of Math. Studies,* Princeton Univ. Press, Princeton, 1992.

[GK9] R. E. Greene and S. G. Krantz, The invariant geometry of bounded domains, preprint.

[GK10] R. E. Greene and S. G. Krantz, Normal families and the semicontinuity of isometry and automorphism groups, *Math. Zeitschrift* 190(1985), 455-467.

[GRS] P. Greiner and E. M. Stein, *Estimates for the $\bar{\partial}$-Neumann Problem,* Princeton University Press, Princeton, 1977.

[GUA] P. Guan, Hölder regularity of subelliptic pseudodifferential operators, *Duke Math. Jour.* 60(1990), 563-598.

[GUN] R. C. Gunning, *Lectures on Complex Analytic Varieties: The Local Parametrization Theorem*, Princeton University Press, Princeton, 1970.

[HAH1] K. T. Hahn, Quantitative Bloch's theorem for certain classes of holomorphic mappings of the ball in $\mathbb{C}P^n$, *J. Reine Angewandte Math.* 283(1976), 99-109.

[HAH2] K. T. Hahn, Asymptotic properties of normal and nonnormal holomorphic functions, *Bull. A.M.S.* 11(1984), 151-154.

[HAS] M. Hakim and N. Sibony, Spectre de $A(\bar{\Omega})$ pour des domaines bornés faiblement pseudoconvexes réguliers, *J. Funct. Anal.* 37(1980), 127-135.

[HAM1] R. Hamilton, Deformation of complex structures on manifolds with boundary. I. The stable case, *J. Diff. Geom.* 12(1977), 1-45.

[HAM2] R. Hamilton, Deformation of complex structures on manifolds with boundary. II. Families of noncoercive boundary value problems, *J. Diff. Geom.* 14(1979), 409-473.

[HAN] Y. S. Han, thesis, Washington University, 1984.

[HJMT] Y. S. Han, B. Jawerth, M. Taibleson, and G. Weiss, Littlewood-Paley theory and ϵ- families of operators, *Colloquium Math.*, LX/LXI(1990), 321-359.

[HSA1] Y. S. Han and E. Sawyer, The weak boundedness property and the *Tb* theorem, *Rev. Math. Iboamericana* 6(1990), 17-41.

[HSA2] Y. S. Han and E. Sawyer, to appear.

[HAG] N. Hanges, Explicit formulas for the Szegö kernel for some domains in $\mathbb{C}^2$, J. Funct. Anal. 88(1990), 153-165.

[HAY] W. Hayman, *Meromorphic Functions*, Oxford University Press, Oxford, 1964.

[HEI1] M. Heins, On the number of 1 - 1 directly conformal maps which a multiply-connected plane region of finite connectivity p (> 2) admits onto itself, *Bull. A.M.S.* 52(1946), 8-21.

[HEI2] M. Heins, On a class of conformal metrics, *Nagoya Math. Journal* 21(1962), 1-60.

[HEL] S. Helgason, *Differential Geometry and Symmetric Spaces*, Academic Press, New York, 1962.

[HIR] M. Hirsch, *Differential Topology*, Springer Verlag, Berlin, 1976.

[HOF] K. Hoffman, *Banach Spaces of Holomorphic Functions*, Prentice-Hall, Englewood Cliffs, 1962.

[HOR1] L. Hörmander, Estimates for translation invariant operators in L^p spaces, *Acta Math.* 104(1960), 93-140.

[HOR2] L. Hörmander, Pseudo-differential operators, *Comm. Pure Appl. Math.* 18(1965), 501-517.

[HOR3] L. Hörmander, Hypoelliptic second order differential equations, *Acta Math.* 119(1967), 147-171.

[HOR4] L. Hörmander, *The Analysis of Linear Partial Differential Operators*, in four volumes, Springer, Berlin, 1983-85.

[HOR5] L. Hörmander, L^p estimates for pluri-subharmonic functions, *Math. Scand.* 20(1967), 65-78.

[HUA1] X. Huang, Some applications of Bell's theorem to weakly pseudoconvex domains, *Pac. Jour. Math.*, to appear.

[HUA2] X. Huang, A preservation principle of extremal mappings near a strongly pseudoconvex point and its applications, *Ill. J. Math.*, to appear.

[HUA3] X. Huang, Boundary behavior of complex geodesics near a strongly pseudoconvex point, *Can. J. Math.*, to appear.

[HUC] A. Huckleberry, lecture, A.M.S. Summer Research Institute, Santa Cruz, 1989.

[HUL] A. Huckleberry and Oeljeklaus, The classification of homogeneous surfaces, *Expositiones Math.* 4(1986), 289-334.

[HUR] W. Hurewicz and Wallman, *Dimension Theory*, Princeton University Press, Princeton, 1948.

[KAN] S. Kaneyuki, *Homogeneous Domains and Siegel Domains*, Springer Lecture Notes #241, Berlin, 1971.

[KAT] Y. Katznelson, *An Introduction to Harmonic Analysis*, John Wiley and Sons, New York, 1970.

[KER] N. Kerzman, The Bergman kernel function. Differentiability at the boundary, *Math. Ann.* 195(1972), 149-158.

[KIM1] K. T. Kim, thesis, UCLA, 1988.

[KIM2] K. T. Kim, On a boundary point repelling automorphism orbits, *Jour. Math. Anal. and Applics.*, to appear.

[KIM3] K. T. Kim, Complete localization of domains with noncompact automorphism groups, *Trans. A. M. S.* 319(1990), 139-153.

[KIM4] K. T. Kim, Domains in $\mathbb{C}^n$ with a Levi flat boundary which possess a noncompact automorphism group, preprint.

[KIM5] K. T. Kim, preprint.

[KLE] P. Klembeck, Kähler metrics of negative curvature, the Bergman metric near the boundary and the Kobayashi metric on smooth bounded strictly pseudoconvex sets, *Indiana Univ. Math. J.* 27(1978), 275-282.

[KNS] A. Knapp and E. M. Stein, Intertwining operators for semisimple grous, *Ann. Math.* 93(1971), 489-578.

[KOB1] S. Kobayashi, The geometry of bounded domains, *Trans. A. M. S.* 92(1959), 267-290.

[KOB2] S. Kobayashi, *Hyperbolic Manifolds and Holomorphic Mappings*, Dekker, New York, 1970.

[KOD1] A. Kodama, On the structure of a bounded domain with a special boundary point, *Osaka Jour. Math.* 23(1986), 271-298.

[KOD2] A. Kodama, On the structure of a bounded domain with a special boundary point II, *Osaka Jour. Math.* 24(1987), 499-519.

[KOD3] A. Kodama, Characterizations of certain weakly pseudoconvex domains $E(k, \alpha)$ in $\mathbb{C}^n$, *Tôhoku Math. J.* 40(1988), 343-365.

[KOD4] A. Kodama, A characterization of certain domains with good boundary points in the sense of Greene-Krantz, *Kodai Math. J.* 12(1989), 257-269.

[KKM] A. Kodama, D. Ma, and S. G. Krantz, A characterization of generalized complex ellipsoids in $\mathbb{C}^n$ and related results, *Indiana Univ. Math. Jour.* 41(1992).

[KON1] J. J. Kohn, Boundary behavior of $\bar{\partial}$ on weakly pseudoconvex manifolds of dimension two, *J. Diff. Geom.* 6(1972), 523-542.

[KON2] J. J. Kohn and L. Nirenberg, On the algebra of pseudo-differential operators, *Comm. Pure Appl. Math.* 18(1965), 269-305.

[KON3] J. J. Kohn, Subellipticity of the $\bar{\partial}$-Neumann problem on pseudoconvex domains: sufficient conditions, *Acta Math.* 142(1979), 79-122.

[KOR1] A. Koranyi, Harmonic functions on Hermitian hyperbolic space, *Trans. A. M. S.* 135(1969), 507-516.

[KOR2] A. Koranyi, Boundary behavior of Poisson integrals on symmetric spaces, *Trans. A.M.S.* 140(1969), 393-409.

[KRA1] S. Krantz, *Function Theory of Several Complex Variables*, 2nd. ed., Wadsworth, Belmont, 1992.

[KRA2] S. Krantz, Lipschitz spaces, smoothness of functions, and approximation theory, *Expositiones Math.* 3(1983), 193-260.

[KRA3] S. Krantz, The boundary behavior of the Kobayashi metric, *Rocky Mt. Jour. Math.* 22(1992), in press.

[KRA4] S. Krantz, On a theorem of Stein I, *Trans. AMS* 320(1990), 625-642.

[KRA5] S. Krantz, Convexity in complex analysis, *Proc. Symp. Pure Math.*, v. 52, Am. Math. Soc., Providence, 1991.

[KRA6] S. Krantz, A compactness principle in complex analysis, *Division de Matematicas, Univ. Autonoma de Madrid Seminarios*, vol. 3, 1987, 171-194.

[KRA7] S. Krantz, *Partial Differential Equations and Complex Analysis*, CRC Press, Boca Raton, 1992.

[KRA8] S. Krantz, Invariant metrics and the boundary behavior of holomorphic functions on domains in $\mathbb{C}^n$, *The Journal of Geometric Analysis* 1(1991), 71-97.

[KRA9] S. Krantz, *Complex Analysis: The Geometric Viewpoint*, a Carus monograph, the Mathematical Association of America, Washington, D. C., 1990.

[KRA10] S. Krantz, Characterizations of smooth domains in $\mathbb{C}$ by their biholomorphic self maps, *Am. Math. Monthly* 90(1983), 555-557.

[KRA11] S. Krantz, Boundary values and estimates for holomorphic functions of several complex variables, *Duke Math. J.* 47(1980), 81-98.

[KRA12] S. Krantz, Holomorphic functions of bounded mean oscillation, *Duke Math. J.* 47(1980), 743-761.

[KRA13] S. Krantz, Fractional integration on Hardy spaces, *Studia Math.* LXXIII(1982), 87-94.

[KRA14] S. Krantz, Characterization of various domains of holomorphy via $\bar{\partial}$ estimates and applications to a problem of Kohn, *Ill. J. Math.* 23(1979), 267-286.

[KRA15] S. Krantz, Geometric Lipschitz spaces and applications to complex function theory and nilpotent groups, *J. Funct. Anal.* 34(1979), 456-471.

[KL1] S. Krantz and Song-Ying Li, A note on Hardy spaces and functions of bounded mean oscillation on domains in $\mathbb{C}^n$, preprint.

[KL2] S. Krantz and Song-Ying Li, On the decomposition theorems for Hardy spaces on domains in $\mathbb{C}^n$ and applications, *Michigan Math. J.*, to appear.

[KRM1] S. Krantz and Daowei Ma, Bloch functions on strongly pseudoconvex domains, *Indiana Univ. Math. Journal* 37(1988), 145-163.

[KRM2] S. Krantz and Daowei Ma, Isometric isomorphisms of the Bloch space, *Michigan Math. Jour.* 36(1989), 173-180.

[KRP1] S. Krantz and H. R. Parks, *A Primer of Real Analytic Functions*, Birkhäuser, Basel, 1992.

[KRP2] S. Krantz and H. R. Parks, *Analysis on Domains*, to appear.

[LAT] R. Latter, A characterization of $H^p(\mathbb{R}^n)$ in terms of atoms, *Studia Math.* 62(1978), 93-101.

[LEV] O. Lehto and K. I. Virtanen, Boundary behavior and normal meromorphic functions, *Acta Math.* 97(1957), 47-65.

[LEM1] L. Lempert, La metrique Kobayashi et las representation des domains sur la boule, *Bull. Soc. Math. France* 109(1981), 427-474.

[LEM2] L. Lempert, private communication.

[LEM3] L. Lempert, Holomorphic retracts and intrinsic metrics in convex domains, *Analysis Mathematica* 8(1982), 257-261.

[LOP] A. J. Lohwater and C. Pommerenke, On normal meromorphic functions, *Ann. Acad. Sci. Fenn. Ser A* I(1973). 12 pp.

[LUQ] Lu Qi-Keng, On Kähler manifolds with constant curvature, *Acta Math. Sinica* 16(1966), 269-281.(Chinese);(=*Chinese Math.* 9(1966)), 283-298.

[MA1] Daowei Ma, thesis, Washington University in St. Louis, 1990.

[MAS] B. Mair and D. Singman, A generalized Fatou theorem, *Trans. A. M. S.* 300(1987), 705-719.

[MPS1] B. Mair, S. Philipp, and D. Singman, A converse Fatou theorem, *Mich. Math. J.* 36(1989), 3-9.

[MPS2] B. Mair, S. Philipp, and D. Singman, A converse Fatour theorem on homogeneous spaces, *Ill. J. Math.* 33(1989), 643-656.

[MCM] D. McMichael, preprint.

[MCN1] J. McNeal, Boundary behavior of the Bergman kernel function in $\mathbb{C}^2$, *Duke Math. J.* 58(1989), 499-512.

[MCN2] J. McNeal, Local geometry of decoupled pseudoconvex domains, *Proceedings of a Conference for Hans Grauert*, Vieweg, 1990.

[MCN3] J. McNeal, Holomorphic sectional curvature of some pseudoconvex domains, *Proc. Am. Math. Soc.* 107(1989), 113-117.

[MCN4] J. McNeal, Lower bounds on the Bergman metric near a point of finite type, preprint.

[MCN5] J. McNeal, Convex domains of finite type, preprint.

[MCN6] J. McNeal, Estimates on the Bergman kernels of convex domains, preprint.

[MER] P. Mercer, thesis, Univ. of Toronto, 1992.

[MIN] D. Minda, Bloch constants, *J. d'Analyse Math.* XLI(1982), 54-84.

[MIY] A. Miyachi, H^p spaces over open subsets of $\mathbb{R}^n$, *Studia Math.* XCV(1990), 205-228.

[MSI] G. D. Mostow and Y. T. Siu, A compact Kähler surface of negative curvature not covered by the ball, *Annals of Math.* 112(1980), 321-360.

[MUN] J. Munkres, *Elementary Differential Topology*, Princeton University Press, Princeton

[NAR] R. Narasimhan, *Several Complex Variables*. Univ. of Chicago Press, Chicago, 1971.

[NRSW] A. Nagel, J. P. Rosay, E. M. Stein, and S. Wainger, Estimates for the Bergman and Szegö kernels in $\mathbb{C}^2$, *Ann. Math.* 129(1989), 113-149.

[NARU] A. Nagel and W. Rudin, Local boundary behavior of bounded holomorphic functions, *Can. Jour. Math.* 30(1978), 583-592.

[NAS1] A. Nagel and E. M. Stein, On certain maximal functions and approach regions, *Advances in Math.* 54(1984), 83-106.

[NAS2] A. Nagel and E. M. Stein, *Lectures on Pseudodifferential Operators*, Princeton University Press, Princeton, 1979.

[NSW1] A. Nagel, E. M. Stein, and S. Wainger, Boundary behavior of functions holomorphic in domains of finite type, *Proc. Nat. Acad. Sci. USA* 78(1981), 6596-6599.

[NSW2] A. Nagel, E. M. Stein, and S. Wainger, Balls and metrics defined by vector fields. I. Basic properties, *Acta Math.* 155(1985), 103-147.

[PAN1] M. Y. Pang, Finsler metrics with properties of the Kobayashi metric on convex domains, preprint.

[PAN2] M. Y. Pang, Smoothness of the Kobayashi metric of non-convex domains, preprint.

[PEE] J. Peetre, Réctifications à l'article "Une caractérisation abstraite des opérateurs différentiels', *Math. Scand.* 8(1960), 116-120.

[PEL] M. Peloso, thesis, Washington University, 1990.

[PHS] D. Phong and E.M. Stein, Estimates for the Bergman and Szegö projections on strongly pseudoconvex domains, *Duke J. of Math.*, 44(1977), 695-704.

[PIN] S. Pinchuk, On the analytic continuation of holomorphic mappings, *Mat. Sb.* 98(140)(1975), 375-392; *Mat. U.S.S.R. Sb.* 27(1975), 416-435.

[POM1] Ch. Pommerenke, *Univalent Functions*, Vandenhoeck and Ruprecht, Göttingen, 1975.

[POM2] Ch. Pommerenke, On Bloch functions, *J. London Math. Soc.* 2(1970), 689-695.

[PRI] I. Priwalow, *Randeigenschaften Analytischer Funktionen*, Deutsch Verlag der Wissenschaften, Berlin, 1956.

[RAMA1] I. Ramadanov, Sur une propriéte de las fonction de Bergman, *C. R. Acad. Bulgare des Sci.* 20(1967), 759-762.

[RAMA2] I. Ramadanov, private communication.

[RAM1] W. Ramey, Boundary behavior of bounded holomorphic functions along maximally complex submanifolds, *Am. J. Math.* 106(1984), 975-999.

[RAM2] W. Ramey, Local boundary behavior of pluriharmonic functions along curves, *Amer. J. Math.* 108(1986), 175-191.

[RAM3] W. Ramey, On the behavior of holomrorphic functions near maximum modulus sets, *Math. Ann.* 276(1986), 137-144.

[REI] H. Reiffen, Die differentialgeometrischen Eigenschaften der invarianten Distanz-funktion von Carathéodory, *Schr. Math. Inst. Univ. Münster*, No. 26(1963).

[RIE] M. Riesz, L'intégrale de Riemann-Liouville et le problème de Cauchy, *Acta Math.* 81(1949), 1-223.

[ROB] A. Robinson, Metamathematical problems, *J. Symbolic Logic* 38(1973), 500-516.

[ROS] J. P. Rosay, Sur une characterization de la boule parmi les domains de $\mathbb{C}^n$ par son groupe d'automorphismes, *Ann. Inst. Four. Grenoble* XXIX(1979), 91-97.

[ROST] L. P. Rothschild and E. M. Stein, Hypoelliptic differential operators and nilpotent groups, *Acta Math.* 137(1976), 247-320.

[ROY] H. Royden, The extension of regular holomorphic maps, *Proc. A.M.S.* 43(1974), 306-310.

[RUT] L. Rubel and R. Timoney, An extremal property of the Bloch space, *Proc. A.M.S.* 75(1979), 45-49.

[ROW] H. Royden and B. Wong, Carathéodory and Kobayashi metrics on convex domains, preprint.

[RUD1] W. Rudin, *Function Theory of the Unit Ball of* $\mathbb{C}^n$, Springer Verlag, Berlin, 1980.

[RUD2] W. Rudin, Holomorphic maps that extend to automorphisms of a ball, *Proc. A.M.S.* 81(1981), 429-432.

[RUD3] W. Rudin, Möbius invariant spaces on balls, *Ann. Inst. Fourier, Grenoble* 33(1983), 19-41.

[SAZ1] R. Saerens and W. Zame, The isometry groups of manifolds and the automorphism groups of domains, *Trans. A. M. S.* 301(1987), 413-429.

[SAZ2] R. Saerens and W. Zame, The local automorphism group of a CR hypersurface, *Proceedings of a Special Year at the Mittag-Leffler Institute*, Ann. of Math. Studies, Princeton University Press, Princeton, 1992.

[SIB1] N. Sibony, unpublished notes.

[SIB2] N. Sibony, A class of hyperbolic manifolds, in *Recent Developments in Several Complex Variables* (J. Fornæss ed.), Ann. of Math. Studies v. 100, Princeton Univ. Press, Princeton, 1981.

[SKW] M. Skwarczynski, The distance in the theory of pseudo-conformal transformations and the Lu Qi-King conjecture, *Proc. A.M.S.* 22(1969), 305-310.

[SMI] K. T. Smith, A generalization of an inequality of Hardy and Littlewood, *Can. Jour. Math.* 8(1956), 157-170.

[STE1] E. M. Stein, *Singular Integrals and Differentiability Properties of Functions*, Princeton University Press, Princeton, 1970.

[STE2] E. M. Stein, Singular integrals and estimates for the Cauchy-Riemann equations, *Bull. A.M.S.* 79(1973), 440-445.

[STE3] E. M. Stein, *Boundary Behavior of Holomorphic Functions of Several Complex Variables*, Princeton University Press, Princeton, 1972.

[STW] E. M. Stein and G. Weiss, On the theory of harmonic functions of several variables I. The theory of H^p spaces, *Acta Math.* 103(1960), 25-62.

[SUE1] J. Sueiro, On maximal functions and Poisson-Szegö integrals, *Trans. A.M.S.* 298(1986), 653-669.

[SUE2] J. Sueiro, A note on maximal operators of Hardy-Littlewood type, *Math. Proc. Camb. Phil. Soc.* 102(1987), 131-134.

[SUE3] J. Sueiro, Tangential boundary limits and exceptional sets for holomorphic functions in Dirichlet-type spaces, *Math. Ann.* 286(1990), 661-678.

[SUY] N. Suita and A. Yamada, On the Lu Qi-Keng conjecture, *Proc. A.M.S.* 59(1976), 222-224.

[SUN] T. Sunada, Holomorphic equivalence problem for bounded Reinhardt domains, *Math. Ann.* 235(1978), 111-128.

[TAW] M. Taibleson and G. Weiss, The molecular characterization of certain

Hardy spaces, *Representation Theorems for Hardy Spaces*, pp. 67-149, *Astérisque* 77, *Soc. Math. France*, Paris, 1980.

[TIM1] R. Timoney, Bloch functions in several complex variables, I, *Bull. London Math. Soc.* 12(1980), 241-267.

[TIM2] R. Timoney, Bloch functions in several complex variables, II, *J. Reine Angew. Math.* 319(1980), 1-22.

[UCH] A. Uchiyama, A constructive proof of the Fefferman-Stein decomposition of $BMO(\mathbb{R}^n)$, *Acta Math.* 148(1982), 215-241.

[ULL] D. Ullrich, personal communication.

[VEL] J. Velling, thesis, Stanford University, 1985.

[VEN] S. Venturini, Intrinsic metrics in complete circular domains, *Math. Annalen*, 288(1990), 473-481.

[VIG1] J. P. Vigue, Sur les points fixes d'applications holomorphes, *C. R. Acad. Sc. Paris* 303(1986), 927-930.

[VIG2] J. P. Vigue, Points fixes d'applications holomorphes dans un domaine borné convex de $\mathbb{C}^n$, *Acta Math.* 166(1991), 1-26.

[VIG3] J. P. Vigue, Points fixes d'une application holomorphe d'un domaine borné dans lui-même, *Acta Math.* 166(1991), 1-26.

[VIT1] A. Vitushkin, Extension of the germ of a holomorphic mapping of a strictly pseudoconvex surface, *Uspekhi Mat. Nauk.* 40(1985), 151-152.

[VIT2] A. Vitushkin, Real-analytic hypersurfaces of complex manifolds, *Uspekhi Mat. Nauk.* 40(1985), 3-31.

[WEB] S. Webster, Biholomorphic mappings and the Bergman kernel off the diagonal, *Invent. Math.* 51(1979), 155-169.

[WEI] G. Weiss, Some problems in the theory of Hardy spaces, *Proc. Symp. Pure Math.*, vol XXXV, part 1, American Mathematical Society, Providence, 1979, 189-200.

[WHI] H. Whitney, Analytic extensions of differentiable functions defined in closed sets, *Transactions of the American Mathematical Society* 36(1934), 63–89.

[WID] D. Widder, *The Heat Equation*, Princeton Univ. Press, Princeton, 1948.

[WIE] J. Wiegerinck, Domains with finite dimensional Bergman space, *Math. Z.* 187(1984), 559-562.

[WONG] B. Wong, Characterizations of the ball in $\mathbb{C}^n$ by its automorphism group, *Invent. Math.* 41(1977), 253-257.

[WU1] H. H. Wu, Normal families of holomorphic mappings, *Acta Math.* 119(1967), 193-233.

[WU2] H. H. Wu, Proceedings of a Special Year at the Mittag-Leffler Institute, *Annals of Math. Studies*, Princeton University Press, Princeton, 1992.

[XIG] Gang, Xiao, On abelian automorphism group of a surface of general type, *Invent. Math.* 102(1990), 619-631.

[YAN] H. Yanagihara, private communication.

[YU1] J. Yu, private communication.

[YU2] J. Yu, Multitypes of convex domains, *Indiana Math. J.* 41(1992), 837-851.

[YU3] J. Yu, Estimates for the Bergman kernel and metric on a convex domain, in preparation.

[YU4] J. Yu, thesis, Washington University, 1993.

[ZAI] M. G. Zaidenberg and V. J. Lin, On bounded domains of holomorphy that are not holomorphically contractible, *Soviet Math. Dokl.* 20(1979), 1262-1266.

[ZAL1] L. Zalcman, A heuristic principle in complex function theory, *Amer. Math. Monthly* 82(1975), 813-817.

[ZAL2] L. Zalcman, Modern perspectives on classical function theory, *Rocky Mt. Jour. Math.* 12(1982), 75-92.

[ZHU1] K. Zhu, Schatten class Hankel operators on the Bergman space of the unit ball, preprint.

[ZHU2] K. Zhu, Analytic Besov spaces, J. Math. Anal. and Applic. 157(1991), 318-336.

[ZHU3] K. Zhu, Hilbert-Schmidt Hankel operators on the Bergman space, *Proc. A. M. S.* 109(1990), 721-730.

[ZHU4] K. Zhu, Möbius invariant Hilbert spaces of holomorphic functions in the ball, *Trans. Am. Math. Soc.* 323(1991), 823-842.

Index

Titles in This Series

(Continued from the front of this publication)

44 **Phillip A. Griffiths,** An introduction to the theory of special divisors on algebraic curves, 1980
43 **William Jaco,** Lectures on three-manifold topology, 1980
42 **Jean Dieudonné,** Special functions and linear representations of Lie groups, 1980
41 **D. J. Newman,** Approximation with rational functions, 1979
40 **Jean Mawhin,** Topological degree methods in nonlinear boundary value problems, 1979
39 **George Lusztig,** Representations of finite Chevalley groups, 1978
38 **Charles Conley,** Isolated invariant sets and the Morse index, 1978
37 **Masayoshi Nagata,** Polynomial rings and affine spaces, 1978
36 **Carl M. Pearcy,** Some recent developments in operator theory, 1978
35 **R. Bowen,** On Axiom A diffeomorphisms, 1978
34 **L. Auslander,** Lecture notes on nil-theta functions, 1977
33 **G. Glauberman,** Factorizations in local subgroups of finite groups, 1977
32 **W. M. Schmidt,** Small fractional parts of polynomials, 1977
31 **R. R. Coifman and G. Weiss,** Transference methods in analysis, 1977
30 **A. Pełczyński,** Banach spaces of analytic functions and absolutely summing operators, 1977
29 **A. Weinstein,** Lectures on symplectic manifolds, 1977
28 **T. A. Chapman,** Lectures on Hilbert cube manifolds, 1976
27 **H. Blaine Lawson, Jr.,** The quantitative theory of foliations, 1977
26 **I. Reiner,** Class groups and Picard groups of group rings and orders, 1976
25 **K. W. Gruenberg,** Relation modules of finite groups, 1976
24 **M. Hochster,** Topics in the homological theory of modules over commutative rings, 1975
23 **M. E. Rudin,** Lectures on set theoretic topology, 1975
22 **O. T. O'Meara,** Lectures on linear groups, 1974
21 **W. Stoll,** Holomorphic functions of finite order in several complex variables, 1974
20 **H. Bass,** Introduction to some methods of algebraic K-theory, 1974
19 **B. Sz.-Nagy,** Unitary dilations of Hilbert space operators and related topics, 1974
18 **A. Friedman,** Differential games, 1974
17 **L. Nirenberg,** Lectures on linear partial differential equations, 1973
16 **J. L. Taylor,** Measure algebras, 1973
15 **R. G. Douglas,** Banach algebra techniques in the theory of Toeplitz operators, 1973
14 **S. Helgason,** Analysis on Lie groups and homogeneous spaces, 1972
13 **M. Rabin,** Automata on infinite objects and Church's problem, 1972
12 **B. Osofsky,** Homological dimensions of modules, 1973
11 **I. Glicksberg,** Recent results on function algebras, 1972
10 **B. Grünbaum,** Arrangements and spreads, 1972
9 **I. N. Herstein,** Notes from a ring theory conference, 1971
8 **P. Hilton,** Lectures in homological algebra, 1971
7 **Y. Matsushima,** Holomorphic vector fields on compact Kähler manifolds, 1971
6 **W. Rudin,** Lectures on the edge-of-the-wedge theorem, 1971
5 **G. W. Whitehead,** Recent advances in homotopy theory, 1970
4 **H. S. M. Coxeter,** Twisted honeycombs, 1970
3 **L. Markus,** Lectures in differentiable dynamics, 1971
2 **G. Baumslag,** Lecture notes on nilpotent groups, 1971